Student's Study Guide & Solutions Manual

Colm Mulcahy

Spelman College

William L. Briggs

University of Colorado at Boulder

Essentials of Using and Understanding Mathematics

A Quantitative Reasoning Approach

Jeffrey O. Bennett

William L. Briggs

Boston San Francisco New York
London Toronto Sydney Tokyo Singapore Madrid
Mexico City Munich Paris Cape Town Hong Kong Montreal

Reproduced by Addison-Wesley from camera-ready copy supplied by the author.

Printed in the United States of America.

ISBN 0-321-10907-4

1 2 3 4 5 6 7 8 9 10 DPC 05 04 03 02

Table of Contents

Introduction

Welcome to the *Study Guide and Solutions Manual* for *Essentials of Using and Understanding Mathematics*. Hopefully, with the help of this book, your course in quantitative reasoning will be both enjoyable and successful. The goal of this guide is not to add to your workload for the course, but to give you a set of concise notes that will make your studying as effective as possible. If you work with this guide as you read and do problems, you should get the most from the course.

This guide is organized according to the units in the textbook. For each unit you will find the following features:

Overview

This section provides a brief survey of the unit, its major points, and its goals. This feature is not a substitute for reading the text!

Key Words and Phrases

This section is simply a list of the important words and phrases used in the unit. You can use this section for review, study, and self-testing. You should be able to explain or define all of the terms on this list.

Key Concepts and Skills

In this section you will find a summary of the most important concepts in each unit. These concepts may be general ideas (for example, the distinction between deductive and inductive arguments) or basic skills (for example, creating the equation of a straight line). This section should also be helpful for review, study, and self-testing.

Important Review Boxes

If a unit has one or more Review Boxes, they are listed. These boxes are quite important for providing background skills or knowledge needed for the unit.

Solutions to (Most) Odd Problems

Following all of the unit summaries, the second half of this Guide provides full solutions to most of the odd-numbered problems in the textbook. Mathematics is not a spectator sport! Reading the solutions is never a substitute for working the problems. *You are strongly advised to work the problems first and then check the solutions.*

How to Succeed In This Course

Using *This* Book

Before we get into more general strategies for studying, here are a few guidelines that will help you use *this* book most effectively.

- Before doing any assigned problems, read assigned material *twice*:
  On the first pass, read quickly to gain a "feel" for the material and concepts presented.
  On the second pass, read the material in more depth, and work through the examples carefully.

- During the second reading, take notes that will help you when you go back to study later. In particular:
  - *Use the margins!* The wide margins in this textbook are designed to give you plenty of room for making notes as you study.
  - Don't highlight — underline! Using a pen or pencil to underline material requires greater care than highlighting, and therefore helps to keep you alert as you study.
- After you complete the reading, and again when studying for exams, make sure you can answer the *review questions* at the end of each unit.
- You'll learn best by *doing*, so work plenty of the end-of-unit problems. Don't be reluctant to work more than the problems that your instructor assigns.

Budgeting Your Time

A general rule of thumb for college classes is that you should expect to study about 2 to 3 hours per week *outside* class for each unit of credit. For example, a student taking 15 credit hours should spend 30 to 45 hours each week studying outside of class. Combined with time in class, this works out to a total of 45 to 60 hours per week — not much more than the time required of a typical job. Moreover, except for class time, you get to choose your own hours. Of course, if you are working while you attend school, you will need to budget your time carefully.

If you find that you are spending fewer hours than these guidelines suggest, you can probably improve your grade by studying more. If you are spending more hours than these guidelines suggest, you may be studying inefficiently; in that case, you should talk to your instructor about how to study more effectively for a mathematics class.

General Strategies for Studying

- Don't miss class. Listening to lectures and participating in discussions is much more effective than reading someone else's notes. Active participation will help you retain what you are learning.
- Budget your time effectively. An hour or two each day is more effective, and far less painful, than studying all night before homework is due or before exams.
- If a concept gives you trouble, do additional reading or problem solving beyond what has been assigned. If you still have trouble, *ask for help*: you surely can find friends, colleagues, or teachers who will be glad to help you learn. Never be reluctant to ask questions or ask for help in this course. If you have a question or problem, it is extremely unlikely that you will be alone!
- Working together with friends can be valuable; you improve your own understanding when discussing concepts with others. However, be sure that you learn *with* your friends and do not become dependent on them.

Preparing for Exams

- Rework problems and other assignments; try additional problems to be sure you understand the concepts. Study your performance on assignments, quizzes, or exams from earlier in the semester.
- Study your notes from lectures and discussions. Pay attention to what your instructor expects you to know for an exam.

- Reread the relevant sections in the textbook, paying special attention to notes you have made in the margins.
- Study individually before joining a study group with friends. Study groups are effective only if *every* individual comes prepared to contribute.
- Don't stay up too late before an exam. Don't eat a big meal within an hour of the exam (thinking is more difficult when blood is being diverted to the digestive system).
- Try to relax before and during the exam. If you have studied effectively, you are capable of doing well. Staying relaxed will help you think clearly.

Finally, good luck! We wish you an enjoyable and rewarding experience in quantitative reasoning.

1 THINKING CRITICALLY

Overview

Before discussing Chapter 1, we urge you to take a few minutes to read the prologue to the textbook. This short chapter sets the stage for the entire book. It presents the idea of quantitative reasoning and discusses the importance of interdisciplinary thinking. It gives a high altitude picture of mathematics and how it impacts many other subjects that you will encounter either in other courses or in your career. Finally, it gives some advice on using the book and studying for your course. It's worth a quick reading. Now on to Chapter 1.

In teaching this course to many students of over many years, we know that often the most serious weakness that students bring to the course is not poor mathematical skills, but poor reasoning skills. Often it's not multiplying two numbers that creates problems, but deciding *when* to multiply! For this reason, the book opens with a chapter that contains virtually no mathematics. The emphasis of the chapter is critical thinking and logical skills.

In this chapter you will encounter some introductory logic, but don't worry; we don't get carried away with symbolic logic and heavy-duty truth tables. In fact, much of this chapter may be familiar to you from previous courses in logic or philosophy.

Unit 1A Recognizing Fallacies

Unit 1A opens the chapter by explaining that a logical **argument** (as opposed to an everyday argument) is a set of facts or assumptions, called **premises**, that lead to a **conclusion**. A **fallacy** is an argument that is either deceptive or wrong. This unit explores common fallacies that you might encounter in advertising or (bad) news reports. We present ten different so-called informal fallacies, some of which may seem quite obvious, others of which may be quite subtle. Critical reading and thinking will help you avoid becoming a victim of these fallacies!

Key Words and Phrases

logic	argument	premise
conclusion	fallacy	appeal to popularity
false cause	appeal to ignorance	hasty generalization
limited choice	appeal to emotion	personal attack
circular reasoning	diversion	straw man

Key Concepts and Skills

- identify the premise and conclusion of an argument.
- recognize informal fallacies in advertisements and news report.

Unit 1B Propositions and Truth Values

Overview

In this unit we introduce formal logic in a very casual way. We start with **propositions** — statements that make a claim that can be true or false. Then we look at the **connectors** that can be used with propositions to make more complex propositions. The connectors that you will encounter are

not (negation)
or (disjunction)
and (conjunction)
if ... then (implications).

Whereas many logic books make heavy use of symbolic logic and truth tables, we will use truth tables primarily for fairly simple propositions that involve one, two, or three connectors. So our excursion into symbolic logic will be limited and designed to provide only an introductory glimpse.

The *if ... then* connector is quite important in both logic and everyday speech (for example, *if I pass this course, then I will graduate*). For this reason, we spend a little time discussing other forms of the proposition *if P, then Q*. These other forms are called the

- **converse** (*if Q, then P*)
- **inverse** (*if not P, then not Q*)
- **contrapositive** (*if not Q, then not P*).

This particular discussion may seem a bit technical, but it's also extremely practical. For example, suppose it's true that *if I read the book, then I will pass the course*. Does it follow that *if I don't read the book, then I won't pass the course*? You will see!

Key Words and Phrases

proposition	truth table	negation
conjunction	disjunction	conditional
antecedent	consequent	converse
inverse	contrapositive	logical equivalence

Key Concepts and Skills

- understand negation, conjunction, disjunction, and conditionals and their truth tables.
- use truth tables to evaluate the truth of compound propositions that use two or more connectors.
- analyze various forms of *if ... then* propositions.

Unit 1C Sets and Venn Diagrams

Overview

You may have encountered Venn diagrams before now as a way to illustrate the relationships between collections of objects, or **sets**. In this unit, we review some of the most basic properties of sets and then discuss how Venn diagram can be used to work with sets. Throughout the unit, the emphasis is on practical applications of sets and Venn diagrams. One of the important applications of Venn diagrams is to illustrate what are called **categorical propositions** of logic. We will study four basic forms of categorical propositions; given a **subject set** *S* and a **predicate set** *P*, the may be related in the following ways:

- All *S* are *P* (for example, all whales are mammals)
- No *S* are *P* (for example, no fish are mammals)
- Some *S* are *P* (for example, some doctors are women)
- Some *S* are not *P* (for example, some teachers are not men).

As we will see, each form of categorical proposition has a specific Venn diagram. Equally important, the negation of each categorical proposition is one of the other propositions in the list above. Specifically, we have the following relations between the four propositions and their negations

Proposition	**Negation**
All *S* are *P*	Some *S* are not *P*
No *S* are *P*	Some *S* are *P*
Some *S* are *P*	No *S* are *P*
Some *S* are not *P*	All *S* are *P*

Key Words and Phrases

set, Venn diagram, subset, disjoint sets, overlapping sets, categorical propositions

Key Concepts and Skills

- use set notation
- construct Venn diagrams for categorical propositions
- put propositions in standard form
- negate categorical propositions
- construct Venn diagrams for three or more sets

Important Review Box

- A Brief Review of Sets of Numbers

Unit 1D Critical Thinking in Everyday Life

Overview

Critical thinking is an approach to problem solving and decision making that involves careful reading (or listening), sharp thinking, logical analysis, good visualization, and healthy skepticism. In this chapter, we present several guidelines design to sharpen critical thinking skills as they apply to practical problems. The following guidelines are all accompanied by one or more specific examples.

- Read or listen carefully
- Look for hidden assumptions
- Identify the real issue
- Use visual aids (pictures, diagrams, tables)
- Understand all the options
- Watch for fine print and misinformation
- Are other conclusions possible?

Key Concepts and Skills

- apply the guidelines of the unit to practical decisions and problems

2 APPROACHES TO PROBLEM SOLVING

Overview

The first chapter of the book was devoted to *qualitative* issues — topics that don't require extensive use of numbers and computation. In this chapter (and the remainder of the book) we turn to *quantitative* matters. Perhaps it's not surprising that we begin our study of quantitative topics with problem solving. The first two units of the chapter deal with a very basic and important problem solving technique, the use of units. The last unit of the chapter is an overview of various problem-solving strategies. It is a valuable chapter whose lessons run through the rest of the book.

Unit 2A The Problem-Solving Power of Units

Overview

Nearly every number that you encounter in the real world is a measure of *something*: 6 billion *people*, 5280 *feet*, 5 trillion *dollars*, 26 *cubic feet*. The quantities that go with numbers are called **units**. And the message of this unit of the book is that using units can simplify problem solving immensely. Indeed the use of units is one of the most basic problem solving tools.

We begin by considering the most basic **simple units** for the fundamental types of measurement. Here are a few examples of simple units.

- length — inches, feet, meters
- weight — pounds, grams
- capacity — quarts, gallons, liters
- time — seconds, hours.

From these simple units, we can build endless **compound units**.

The rest of the unit is spent illustrating a wide variety of such compound units. Among the many you will meet and work with are units of

- units of area such as square feet and square yards
- units of volume such as cubic inches and cubic feet
- units of speed such as miles per hour
- units of price such as dollars per pound
- units of gas mileage such as miles per gallon.

One of the realities of life is that there are many units for the same quantity. For example, we can measure lengths in inches, centimeters, feet, meters, miles, or kilometers. For example, we measure rooms in square feet, but have to buy carpet in square yards. Thus one of the necessities of problem solving is being able to convert from one unit to another consistent unit. The key to doing conversions between units is to realize that there are three equivalent ways to express a conversion factor. For example, we can say 1 foot = 12 inches, or we can say 1 foot *per* 12 inches, or we can say 12 inches *per* foot. Mathematically we can write

$$1 \text{ foot} = 12 \text{ inches} \quad \text{or} \quad \frac{1 \text{ foot}}{12 \text{ inches}} \quad \text{or} \quad \frac{12 \text{ inches}}{1 \text{ foot}}.$$

These three forms of the same conversion factor are absolutely equivalent. The key to a happy life with units is choosing the appropriate form of the conversion factor for a given situation.

Additional problem solving skills can then be built on these fundamental skills. You will see and learn how to make a chain of conversions factors to solve more complex problems. For example, the time required to count a billion dollars at the rate of one dollar per second is

$$\$1{,}000{,}000{,}000 \times \frac{1 \text{ sec}}{\$1} \times \frac{1 \text{ min}}{60 \text{ sec}} \times \frac{1 \text{ hr}}{60 \text{ min}} \times \frac{1 \text{ day}}{24 \text{ hr}} \times \frac{1 \text{ yr}}{365 \text{ days}} = 31.7 \text{ years.}$$

The fact that the units cancel and give an answer in *years* tells you that the problem has been set up correctly.

Finally we look at a very practical type of unit conversion problem, those associated with currency. If you travel to France you will quickly learn that 1 franc is equal to about 18 cents. From this fact, you will also want to answer questions such as

- which is larger, 1 franc or 1 dollar?
- how many francs in a dollar?
- how many dollars in a franc?
- how many dollars in a 23.45 francs?
- if apples cost 23 francs per kilogram, what is the price in dollars per pound?

Here are a few final words of advice: Never was the motto *practice makes perfect* more true than with problem solving and units. You should work all assigned problems, and then some, in order to master these techniques. And unless your instructor tells you otherwise, there is no need to memorize hundreds of conversion factors. It's helpful to know a few essential conversion factors off the top of your head. As for the rest, it's easiest just to know how to find them quickly in the book.

Key Words and Phrases

simple units	compound units	area
volume	conversion factor	

Key Concepts and Skills

- convert from one simple unit to another; for example, from inches to yards.
- convert from one unit of area to another unit of area; for example, from square inches to square yards.
- convert from one unit of volume to another unit of area; for example, from cubic inches to cubic yards.
- solve problems involving chains of conversion factors; for example, finding the number of seconds in a year.
- convert from one unit of currency to another; for example, from dollars to pesos.

Important Review Box

- A Brief Review of Working with Fractions

Unit 2B Standardized Units: More Problem-Solving Power

Overview

In this unit we continue the study of units and conversion factors, but now look more deeply into the two standard systems of units: The U.S. Customary System of Measurement (or USCS system, which is used primarily in the Unites States) and the metric system (which is used everywhere else in the world).

We first proceed systematically and survey the USCS units for length, weight, and capacity. Tables 2.2, 2.3, and 2.4 contain many conversion factors, but you should focus on *using* these conversion factors, not memorizing them! Having seen the complications of the USCS system, the metric system should come as a welcome relief. Next we present the metric units for length, weight and capacity. As you will see, the system is based on powers of ten and standard prefixes, which makes conversions between units relatively simple.

Unfortunately, for people living in the United States, conversions between the metric system and the USCS system are necessary. (If you would like to avoid doing such problems, then you should work on getting the United States to go metric!) Table 2.6 has a few of the essential conversion factors between the metric and USCS systems. You may want to write other useful conversion factors in the margin near this table.

Finally, we explore two more general families of units. **Energy** (what makes things move or heat up) and **power** (that rate at which energy is used) are incredibly important concepts in understanding the world around us. Uses of energy and power units are presents in matters such as utility bills, diet, and the environment. The other category of units is **density** and **concentration**. In measuring population density, the capacity of computer discs, levels of pollution, and blood alcohol content, these units are indispensable.

Armed with all of these units and conversion factors (and don't forget currency conversion factors as well), we can do even more elaborate problem solving; this is the goal of the remainder of the unit and the problems at the end of the unit.

Key Words and Phrases

USCS system	metric system	meter
gram	liter	second
metric prefixes	Celsius	Fahrenheit
Kelvin	energy	power
kilowatt-hour	watt	kilowatt-hour
density	population density	concentration

Key Concepts and Skills

- given the required conversion factor, convert between two consistent USCS units; for example, rods to miles.
- multiply and divide powers of ten.
- know basic metric units and commonly used prefixes.
- given the required conversion factor, convert between two consistent metric units; for example, millimeters to kilometers.
- given the required conversion factor, convert between a USCS unit and a consistent metric unit; for example, ounces to liters.
- convert between temperatures in the three standard systems.
- solve problems using chains of conversion factors involving USCS units, metric units, and currency units.
- understand and apply units of energy and power.
- understand and apply units of density to materials, population, and information.
- understand and apply units of concentration to pollution, and blood alcohol content.

Important Review Box

- A Brief Review of Powers of Ten

Unit 2C Problem Solving Guidelines and Hints

Overview

Having seen very specific examples of problem solving in the previous two units, we now present some strategies for problem solving in general. *There is no simple and universal formula for solving all problems*! Only continual (and hopefully enjoyable) practice can led one towards mastery in problem solving. This unit is designed to provide you some of that practice.

The unit opens with a very well known four-step process for approaching problem solving. As you will see, this process is not a magic formula, but rather a set of guidelines. The four basic steps are

- understand the problem,
- devise a strategy,
- carry out the strategy, and
- look back, check, interpret and explain your solution.

The remainder of the unit is a list of eight strategic hints for problem solving. Here are the strategic hints:

- there may be more than one answer.
- there may be more than one strategy.
- use appropriate tools.
- consider simpler, similar problems.
- consider equivalent problems with simpler solutions.
- approximations can be useful.
- try alternative patterns of thought.
- don't spin your wheels.

Most of the unit consists of examples in which the four-step process and these strategic hints are put to use. Study these problems and solutions, and try to see how the techniques might be used in other problems that you might encounter. Most important of all, try to enjoy problem solving!

Key Concepts and Skills

- carry out the four-step problem solving process.
- identify and use the eight strategic problem solving hints.

---

3 NUMBERS IN THE REAL WORLD

Overview

In this chapter we explore how numbers are used in real and relevant problems. The first unit deals with percentages and is arguably one of the most practical and important units of the book. As you know, many real world numbers are incredibly large (the federal debt or the storage capacity of a computer disk) or very small (the diameter of a cancer cell or the wavelength of an x-ray); so in the next unit we introduce scientific notation to deal with large and small numbers. The next unit discusses another reality of numbers in the real world: they are often approximate or subject to errors. Finally, the last unit is a fascinating exploration of how numbers deceive us. All in all it's a very practical and useful chapter.

Unit 3A Uses and Abuses of Percentages

Overview

If you read a news article, a financial statement, or an economic report, you will see that one of the most common ways to communicate quantitative information is with percentages. In this units we will explore the many ways that percentages are used — and abused. Percentages are used for three basic purposes:

- as fractions (for example, 45% of the voters favored the incumbent),
- to describe change (for example, taxes increased by 10%), and
- for comparison (for example, women live 3% longer than men).

In this unit, we examine each of these uses of percentages in considerable detail with plenty of examples taken from the news and from real situations. The use of percentages as fractions is probably familiar. The use of percentages to describe change relies on the notions of **absolute change** and **relative** (or **percentage change**). When some quantity changes from a previous value to a new value, we can define its changes in either of two ways:

$$\text{absolute change} = \text{new value} - \text{previous value}$$

$$\text{relative change} = \frac{\text{absolute change}}{\text{previous value}} = \frac{\text{new value} - \text{previous value}}{\text{previous value}}.$$

The use of percentages for comparison relies on the notions of **absolute difference** and **relative** (or **percentage difference**). To make comparisons between two quantities, we must identify the **compared quantity** (the quantity that we are comparing) and the **reference quantity** (the quantity that we are comparing to). For example, if we ask how much larger is one meter than one yard, *one meter* is the compared quantity and *one yard* is the reference quantity. Then we can calculate two kinds of difference:

$$\textbf{absolute difference} = \text{compared quantity} - \text{reference quantity}$$

$$\textbf{relative difference} = \frac{\text{absolute difference}}{\text{reference quantity}} = \frac{\text{compared quantity} - \text{reference quantity}}{\text{reference quantity}}$$

A powerful rule for interpreting statements involving percentages is what we call the **Of Versus More Than Rule**. It says that if the compared quantity is *P% more than* the reference quantity, then it is (100 + *P*)% *of* the reference quantity. For example, if Bess' salary is 30% more than Bob's salary, then Bess' salary is 130% of Bob's salary. Similarly, if Jill's height is 30% *less than* Jack's height, then Jill's height is 70% of Jack's height.

All of these ideas are assembled in this unit to solve a variety of practical problems. Our experience in teaching this subject is that the greatest difficulty is understanding the problem and translating it into mathematical terms. It is important to read the problem carefully, draw a picture if necessary, decide how percentages are used in the problem (as a fraction, for change, or for comparison), and to write a mathematical sentence that describes the situation. It is important to study the examples in the unit and work plenty of practice problems.

The unit closes with several examples of ways that percentages are abused. Be sure you can identify problems in which the previous value shifts and problems in which the quantity of interest is itself a percentage.

Key Words and Phrases

absolute change relative change percentage change

reference quantity	compared quantity	absolute difference
relative difference	percentage difference	of versus more than rule

Key Concepts and Skills

- identify and solve problems that use percentages as fractions.
- identify and solve problems that use percentages to describe change.
- identify and solve problems that use percentages for comparison.
- apply the *of versus more than rule* to practical problems.
- identify fallacies in statements involving percentages.

Important Review Boxes

- A Brief Review of Percentages
- A Brief Review of What is a Ratio?

Unit 3B Putting Numbers in Perspective

Overview

Numbers in the world around us are often very large or very small. In order to write large and small numbers compactly, without a lot of zeros, we use **scientific notation**. This is the main new mathematical idea in this unit. If you haven't used scientific notation before, you will want to practice writing numbers in scientific notation, and multiplying and dividing numbers in scientific notation.

We next spend some time showing how to make rough calculations using estimation. Often we can use approximate values of quantities and arrive at useful **order of magnitude** answers.

All of us suffer from number numbness: large and small numbers eventually lose all meaning. For example, most of us have no sense of how large $5 trillion (the federal debt) or 6 billion (the world's population) really are. Therefore, just as important as *writing* large and small numbers is *visualizing* large and small numbers. We look at some fascinating methods, such as **scaling**, for giving large and small numbers meaning. The goal is often to associate numbers with a striking visual image. For example, if the Earth were the size of a ballpoint pen tip, the Sun would be a grapefruit 15 meters away.

Key Words and Phrases

scientific notation	order of magnitude	scaling
scale ratio	light-year	

Key Concepts and Skills

- write numbers in scientific notation.
- multiply and divide numbers in scientific notation.
- use estimates to compute order of magnitude answers.
- use scaling methods for visualizing large and small numbers.

Important Review Boxes

- A Brief Review of Working with Scientific Notation

Unit 3C Dealing With Uncertainty

Overview

While they may appear totally reliable, the numbers we see in the news or in reports often are quite uncertain and prone to errors. This unit is reminds us of this fact in many different ways, as we consider the sources of errors and the measurement of uncertainty in numbers.

The precision of a number can be described by the number of **significant digits** it has. Significant digits are those digits in a number that we can assume to be reliable, although, this is where care must be used when reading numbers. It's usually easy to determine the number of significant digits in a number; there are a few subtle cases that are listed in the summary box in the unit.

Errors enter problems and calculation in two ways: Errors that arise in unpredictable and unavoidable ways are called **random errors**. On the other hand, **systematic errors** arise due to a problem in the system that affects all measurements in the same way; these errors can often be avoided.

We next discuss **absolute errors** and **relative errors**. It is important to note the analogy between absolute/relative errors and absolute/relative difference.

The next observation in the unit concerns the difference between **accuracy** and **precision**. A pharmacist's scale that can weigh fractions of an ounce has much more *precision* that a butcher's scale that can measure only in pounds. Thus precision refers to how precisely a quantity can be measured or how precisely a number is reported.

On the other hand the *accuracy* of a measurement refers to how close it is to the exact value (which we often don't know). Suppose a marble weighs 4.5 ounces. If one scale reports a weight of 4.4 ounces and another scale reports a weight of 4.7 ounces, the first measurement is more accurate; both measurements have the same precision — to the nearest tenth of an ounce.

A few technicalities arise when it comes to doing arithmetic with approximate numbers. There are two rules that tell us how much precision should be assigned to the sum, difference, product, or quotient of two approximate numbers.

- when adding or subtracting two approximate numbers, the result should be rounded to the same precision as the *least precise* number in the problem.
- when multiplying or dividing two approximate numbers, the result should be rounded to the same number of significant digits as the number in the problem with the *fewest significant digits*.

Key Words and Phrases

significant digits	random errors	systematic errors
absolute error	relative error	accuracy
precision		

Key Concepts and Skills

- determine the number of significant digits in a given number
- distinguish between random errors and systematic errors
- compute absolute and relative errors
- understand the difference between accuracy and precision
- determine the precision of the sum or difference of two approximate numbers
- determine the precision of the product or quotient of two approximate numbers.

Important Review Boxes

- A Brief Review of Rounding

Unit 3D How Numbers Deceive: Polygraphs, Mammograms, and More

Overview

The topics in this unit may seem a bit unrelated, but they are connected by the common theme of how numbers can be deceptive. The unit begins with a curious phenomenon known as **Simpson's paradox**. It can occur in many different ways, but it always arises when quantities are averaged and some crucial information is missing. The examples illustrate the effect in many ways and should be studied carefully.

We next turn to surprisingly deceptive results that occur when using percentages in what are called *two-by-two tables* (such as Table 3.4). These tables are useful to describe two different outcomes (for example, cured and not cured) for two different groups of people (for example, treatment and no treatment). Situations such as this give rise to the terms **false positive, false negative, true positive, and true negative**. Several practical examples are given as they apply to medical tests and lie detector and drug tests. This is a very practical unit that contains some very subtle ideas.

Key Words and Phrases

Simpson's paradox	false positive	false negative
true positive	true negative	

Key Concepts and Skills

- identify and explain Simpson's paradox.
- compute percentages in tables and interpret them correctly.
- understand the subtleties in interpreting medical, drug, and polygraph tests.

4 FINANCIAL MANAGEMENT

Overview

One of the most immediate ways in which mathematics affects every person's everyday life is through finances: bank accounts, credit cards, loans, and investments. In this chapter we take an in-depth look at personal financial matters. As you will see there are many relevant topics in this chapter, and plenty of good mathematics.

Unit 4A The Power of Compounding

Overview

The phenomenon of **compounding** plays a fundamental role in all of finance, both on the investment side of the coin (earning money) and on the loan side (borrowing money). The unit opens by considering simple and compound interest problems as they arise in banking. Compound interest problems rely on four pieces of information:

- **initial deposit**, which we call P,
- **annual percentage interest rate**, which we call APR,
- **number of compoundings per year**, which we call n, and

- **number of years** the account is held, which we call *Y*.

You will see that there are precise formulae that tell you how much a bank account increases in value if you know these four pieces of information. These formulae (and others in the chapter) can get rather complicated and must be evaluated on a calculator. Be sure you study the highlighted boxes that show you how to use your calculator on these formulae; needless to say, practice helps immensely!

An important distinction must be made between the *APR* and the **annual percentage yield** (*APY*) of a bank account. If an account uses compounding at more than once a year, then the balance will increase by *more* than the *APR* in one year (due to the power of compounding). The amount by which the balance in the account actually increases in a year is the *APY*.

As you will see, the more often compounding takes place during the year, the more the balance increases. The limiting case occurs when compounding takes place every instant, or **continuously**. With continuous compounding, you get the maximum return on your money (for a fixed *APR*). The continuous compounding formula involves the mathematical constant *e*, which is approximately 2.71828. You should become familiar with how to compute with *e* on your particular calculator.

The unit concludes with a very practical problem. The usual compound interest formulae tell you how much money you will have in your account after *Y* years with a given initial deposit *today*. But what if you know you would like to have, say $30,000, in 20 years? How much should you deposit *today* in order to reach this goal? This is an example of a present value problem value, and it's quite important for planning purposes.

Key Words and Phrases

simple interest	principal	compound interest
annual percentage rate (APR)	compound interest formula	annual percentage yield (APY)
continuous compounding		

Key Concepts and Skills

- determine the balance in an account with simple interest with a given initial deposit and interest rate.
- determine the balance in an account with compounding with a given initial deposit, *APR*, number of compoundings, and number of years.
- determine the balance in an account with continuous compounding with a given initial deposit, *APR*, and number of years.
- understand the difference between *APR* and *APY,* and be able to compute the *APY* for an account.
- determine the present value for an account that will produce a given balance after a specified number of years.

Important Review Boxes

- A Brief Review of Three Basic Rules of Algebra

Unit 4B Savings Plans

Overview

An **annuity** is a special kind of savings plan in which deposits are made regularly, perhaps every month or every year. People who are creating a college fund or building a retirement plan usually use an annuity. An annuity account increases in value due to compounding (as in regular bank accounts as studied in the previous unit) and due to the regular deposits. Not surprisingly, the mathematics of annuity plans follows naturally from the compound interest formulae of the previous unit.

There is just one basic formula that needs to be mastered in order to work with annuities. The formula requires all of the input needed for the compound interest formula of the previous unit (initial deposit, APR, number of compoundings, and number of years), *plus* the amount of the regular deposits. To keep matters simple (but still realistic) we assume that the regular deposits are made as often as interest is compounded. There is no doubt that the savings plan formula is complicated. Be sure to study the *Using Your Calculator* box to learn how your calculator can be used to evaluate the formula.

Just as with a bank account, it is practical to ask present value questions with savings plans. If you know how much money you would like to have at a future time (perhaps the time of retirement), then how much should you deposit, say monthly, between now and then in order to reach that goal? You will see some practical examples of solving present value problems.

Key Words and Phrases

savings plan	annuity	savings plan formula
total return	annual returns	

Key Concepts and Skills

- determine the value of a savings plan given the initial deposit, APR, number of compoundings, number of years, and the amount of the regular deposits
- determine the present value for a savings plan that will yield a given balance after a specified number of years.
- determine the annual yield on an investment given the total return.

Important Review Boxes

- A Brief Review of Algebra with Powers and Roots

Unit 4C Loan Payments, Credit Cards, and Mortgages

Overview

While most people have bank accounts or investments that earn money, the sad reality is that most people also have debts due to loans of one kind or another. In this unit we will explore the most common kinds of loans: short-term loans (such as automobile loans), loans on credit cards, and **mortgages** (or house loans).

One basic formula governs all loan problems; it is called the **loan payment formula**. Given

- the amount of the loan (called the **principal**),
- the interest rate on the loan (still called the *APR*),
- the number of payment periods per year (usually taken to be 12 for monthly payments), and
- the term of the loan (the number of years over which it will be paid back),

the loan payment formula tells you the amount of the regular payments on the loan. Once again, the loan payment formula is rather complicated, so be sure to study the *Using Your Calculator* box.

The entire unit is devoted to using the loan payment formula for various practical problems. There are a lot of strategic questions involved in choosing a loan. Often you must choose between a loan with a high interest rate and a short term and a loan with a lower interest rate but a longer term. Which is the best choice? You will see several examples of such decisions.

The biggest loan that many people will every use is a house loan or mortgage. Because mortgages involve such large amounts of money over long periods of time, the decision you make in choosing a mortgage are critical. We explore some of the options and strategies involved with mortgages. Specifically, we explain the differences and advantages of fixed-rate mortgages as opposed to adjustable rate mortgages. We also discuss issues of prepayments, refinancing, points, and closing costs. All in all, you will find this to be quite a practical unit, one whose lessons you probably will need someday!

Key Words and Phrases

loan principal	installment loan	loan payment formula
mortgage	down payment	fixed rate mortgage
adjustable rate mortgage	closing costs	points
prepayment penalties	refinance	

Key Concepts and Skills

- know the terminology associated with loans.
- determine the loan payment given the principal, the APR, the number of payment periods, and the term of the loan.
- analyze two loan options to determine which is best for a given situation.
- be familiar with the practical issues associated with a home mortgage, such as points, closing costs, and fixed rate vs. adjustable rate loans.

---

5 STATISTICAL REASONING

Overview

Much of the quantitative information that flows over us every day is in the form of data that is gathered through surveys and other statistical studies. When a person, business, or organization wants to know something, often the first (and only) idea that comes to mind is "collect data." From news reports to scientific research, from political polls to TV viewer surveys, we are surrounded by statistics. How are these numbers gathered? Are the conclusions based on those numbers reliable? And most important of all, should you believe the results of a statistical study? Should you change you life based on the results of a statistical study?

These are the questions that we will *begin* to answer in this chapter. The emphasis of this chapter is on qualitative issues, and there will not be a lot of computation involved. If you master this chapter, you will have a good foundation in statistical studies — enough to allow you to read the news critically and to take a complete course in statistics. All in all, this is a fascinating chapter, filled with many practical and real examples. It has an immediate relevance to your studies and your everyday life.

Unit 5A Fundamentals of Statistics

Overview

The goal of this unit is to learn how a statistical study should (and should *not*) be done. Of utmost importance is the distinction between the **population**, the group of people or objects that you would like to learn about, and the **sample**, the group of people and objects that you actually measure or survey. As you will see, there are basically two aspects to any statistical study:

- collecting data from the people or objects in the sample, and
- drawing conclusions about the entire population based on the information gathered from the sample.

There are many ways that either of these steps can go wrong. Most of this chapter will deal with the first step, which is often called **sampling**. This step is crucial; if a sample is not chosen so that it is representative of the entire population, the conclusions of the study cannot be reliable.

Four different sampling methods will be discussed:

- simple random sampling
- systematic sampling
- convenience sampling
- stratified sampling

It is also important to distinguish between two basic types of statistical studies. An **observational study** is one in which participants (people or objects) are questioned, observed, or measured, but in no way manipulated. By contrast, in an **experiment**, the participants are manipulated in some way, perhaps dividing them into a **treatment group** and a **control group**. A subtle borderline situation is a **case-control study**, in which participants naturally form two or more groups by choice (for example, people who choose to smoke and not to smoke). While a case-control study looks like an experiment, it is actually an observational study.

Other features of statistical studies that should be noticed are the **placebo effect** (when people who *think* they received a treatment respond as if they received the treatment, even though they did not), **single blinding** (when participants do not know whether they receive a treatment), and **double blinding** (when neither researchers nor participants know who received a treatment).

You often hear statements like, "The President's approval rating is 54% with a margin of error of 5%." Results of surveys are often given in this form; for this reason, we will also look briefly the notions of **confidence interval** and **margin of error**. Although these ideas are rather technical, it is important to understand at least what they mean. In Chapter 6, we will see how confidence intervals and margins or error are actually found.

Key Words and Phrases

sample	population	raw data
sample statistics	population parameters	representative sample
simple random sampling	systematic sampling	convenience sampling
stratified sampling	bias	observational study
experiment	treatment group	control group
placebo	placebo effect	single blind experiment
double-blind experiment	case control study	cases
controls	margin of error	confidence interval

Key Concepts and Skills

- understand the distinction between a sample and the population
- know the five basic steps of a statistical study
- be able to form a representative sample by various sampling methods
- understand the difference between an observational study, an experiment, and a case-control study
- understand the placebo effect, and single and double blinding
- interpret margins of error and confidence intervals

Unit 5B Should You Believe a Statistical Study?

Overview

Suppose you read that 68% of all TV viewers watched the NCAA basketball final game or that 1 in 5 people in the world do not have access to fresh drinking water, or that a pre-election poll says that the Republican candidate leads by 5 percentage points. In this unit, we address the basic question: how do you know whether to believe the claims of these statistical studies?

The unit takes the form of a list of eight guidelines for evaluating a statistical study. Each guideline is accompanied by examples and analyses. It may not be possible to assess a particular study in light of *all* eight guidelines, but the more of these guidelines that a study satisfies, the more confident you can be about the conclusions of the study. Just for completeness, we'll list the eight guidelines here.

1. Identify the goal, type and population of the study.
2. Consider the source.
3. Look for bias in the sample.
4. Look for problems in *defining* or *measuring* the quantities of interest.
5. Watch for confounding variables.
6. Consider its setting and wording in surveys.
7. Check that results are presented fairly.
8. Stand back and consider the conclusions.

Key Words and Phrases

bias	variables	selection bias
participation bias		

Key Concepts and Skills

- understand the eight guidelines for evaluating a statistical study, and be able to carry them out on a particular study.

Unit 5C Statistical Tables and Graphs

Overview

We often think of data and statistics as long lists of numbers, and indeed they start in that form. But these numbers don't become meaningful until they are summarized in some digestible form; and one of the best ways to represent long lists of numbers is with a picture. In this and the next unit, we will explore the many ways in which quantitative information can be displayed. This unit focuses on the basic types of graphs, ones that can often be drawn with a pencil and paper. The next unit highlights the more exotic types of graphs which are more difficult to produce, but equally common in news and research reports.

Often the best first step in summarizing data is to make a **frequency table** that shows how often each category in a study appears (for example, the number of students receiving grades of A, B, C, D, and F on an exam). Such a table can also show the **relative frequency** and the **cumulative frequency** for each category.

The type of display used for a set of data also depends on the type of data that are collected. We distinguish between **qualitative data** and **quantitative data**.

Here is a summary of the types of graphs that will be considered in this unit.

- **Bar graphs** are used to show how some numerical quantity (for example, population) varies from one category to another (for example, for several different countries). The categories are usually qualitative and can be shown in any order.
- **Pie charts** are used to show the fraction (or percentage) of a population that falls into various categories. The categories are usually qualitative and can be shown in any order. For example, a pie chart could be used to show the fraction of a class has brown, black, blond, and red hair.
- **Histograms** are bar graphs in which the categories are quantitative, and thus have a natural order. Histograms often show how many objects or people are in various categories. For example, a histogram would be used to show the number of people in a town that fall into age categories 0–9, 10–19, 20–29, and so on.
- **Line charts** serve the same purpose as histograms, but instead of using bars to indicate the number of objects or people in each category, they use dots connected by lines.
- **Time-series diagrams** are usually histograms of line charts that show how a quantity changes in time. For example, the day-to-day changes in the stock market would be displayed as a time-series diagram.

We might point out one possible source of confusion. There seems to be no standard definition of histograms and bar graphs. We have tried to choose a distinction that is common in many books: a histogram is any graphs that uses bars, while a histogram is a particular kind of bar graph that is used for ordered quantitative data.

Key Words and Phrases

bar graph	pie chart	histogram
line chart	time-series diagram	scatter plot

Key Concepts and Skills

- construct a frequency table for a set of table showing frequencies, relative frequencies, and cumulative frequencies.
- determine an appropriate kind of display for a given set of data.
- display an appropriate set of data with a bar graph by finding the correct height of the bars.
- display an appropriate set of data with a pie chart by finding the correct angles for the sectors of the pie.
- display an appropriate set of data with a histogram by finding the correct height of the bars.
- display an appropriate set of data with a line chart by finding the correct location of points on the line chart.
- display an appropriate set of data with a time-series diagram by finding the correct location of points on the diagram.
- display an appropriate set of data with a scatter plot by finding the correct location of points on the plot.

Unit 5D Graphics in the Media

Overview

Whereas the previous unit dealt with graphs that are relatively easy to produce yourself, this unit will survey more complicated displays that appear frequently in the media. The goal in this unit is not to actually produce these more sophisticated graphs; often, powerful software packages are needed to create them. Rather the emphasis will be on interpreting these displays of quantitative information.

Here is a summary of the types of graphs that we will consider in this unit.

- **Multiple bar graphs** are used to show how *two or more* numerical quantities (for example, population and birth rate) vary from one category to another (for example, for several different countries). Multiple bar graphs are really two or more bar graphs combined with each other; they include multiple histograms.
- **Stack plots** are used to display several quantities simultaneously often showing how they all change in time. For example, a stack plot could be used to show how the incidence of several diseases have change in time.
- **Geographical plots**, such contour plots and weather maps, give a two-dimensional picture of how one quantity varies over a geographical region. Such displays are also used to show the distribution of a disease.
- **Three-dimensional** graphs take many different forms, but they are all used to show how three different quantities (or variables) are related to each other.

As graphical displays become more complex, there is more opportunity for confusion and deception. The unit closes by discussing several ways in which you can be misled by such displays. Although graphs can be beautiful and effective, they must be interpreted with caution!

Key Words and Phrases

multiple bar graph	stack plot	three-dimensional graphics
contour plots	exponential scale	pictograph
perceptual distortions	percent change graphs	

Key Concepts and Skills

- interpret multiple bar graphs and histograms
- interpret stack plots
- interpret three-dimensional graphics
- interpret contour plots
- detect deceptive pie charts
- interpret percent change graphs
- interpret exponential scales
- detect deceptive pictographs.

Unit 5E Correlation and Causality

Overview

This unit deals with the critical task of determining whether one event *causes* another event. The process often begins by looking for **correlations** between two variables; a correlation exists when higher values of one variable are consistently associated with higher (or lower) values of the second variable. For example, body weight is correlated with height, because in general, higher body weights mean greater height. Correlations can be detected by making a **scatter diagram** of the two variables in question.

There are three possible explanations for a correlation between two variables. It may be due to

- a coincidence,
- a common underlying cause, or
- a genuine cause and effect relation.

Of the three possibilities, the most interesting and the most difficult to establish is the last one: when can we be sure that one event causes another?

The unit presents six methods (attributed to the philosopher John Stuart Mill) for establishing cause and effect relations:

1. Look for situations in which the effect is correlated with the suspected cause even while other factors vary.

2. Among groups that differ only in the presence or absence of the suspected cause, check that the effect is similarly present or absent.

3. Look for evidence that larger amounts of the suspected cause produce larger amounts of the effect.

4. If the effect might be produced by other potential causes (besides the suspected cause), make sure that the effect still remains after accounting for these other potential causes.

5. If possible, test the suspected cause with an experiment. If the experiment cannot be performed with humans for ethical reasons, consider doing the experiment with animals, cell cultures, or computer models.

6. Try to determine the physical mechanism by which the suspected cause produces the effect.

Because causality is important ways in legal cases, we also look at the legal standards for establishing causality. They are crucial in determining the guilt or innocence of defendants. They are

- possible cause,
- probable cause, and
- cause beyond reasonable doubt.

The unit also presents three (real) case studies in which identifying a cause and effect relation led to a discovery or the solving of a mystery.

Key Words and Phrases

correlation	scatter diagram	positive correlation
negative correlation	coincidence	underlying cause
causality	probable cause	possible cause
beyond reasonable doubt		

Key Concepts and Skills

- construct a scatter diagram for two variables
- identify positive, negative, and no correlation between two variables
- understand the three explanations for a correlation
- apply the six methods for determining causality
- understand the three legal levels of confidence in causality

Unit 5F Characterizing a Data Distribution

Overview

Typical data sets often contain hundreds or thousands of numbers, so one goal is to summarize data sets in compact and meaningful ways. In this unit we explore methods used to describe and summarize data beyond the frequency tables discussed in the previous chapter.

For a very concise summary of the data, it is common to compute the **mean**, **median**, and/or **mode** of the data. These one-number summaries are defined as follows:

- The *mean* of a data set is calculated by the formula
$$\text{mean} = \frac{\text{sum of all values}}{\text{total number of values}}.$$
- The *median* is the middle score in the data set. Note that there will be two "middle" values if a data set has an even number of data points; if the two middle values are different, the median lies halfway between them.
- The *mode* is the most common score in a data set. A data set may have more than one mode, or no mode.

Having determined where the "center" of the data set lies, it is next useful to describe the shape of the data: are the data values symmetric? Are they weighted to the right or left? These questions can be answered qualitatively by finding whether the data values are **positively** or **negatively skewed**.

All of the above ideas can be captured in one compact description of the data called the **five-number summary**, which consists of the **median**, the **upper** and **lower quartiles**, and the **highest** and **lowest data value**. The five-number summary can be displayed nicely with a **box plot**.

More insight into a data set can be gained when we quantify the spread or **dispersion** of the data. The key quantities for this purpose are the **variance** and the **standard deviation**. You will see detailed calculations of these quantities.

All of these methods and measures are accompanied by numerical examples in the text. Be sure you understand how these calculations are done and then practice on the data sets in the problems. It is also helpful to learn your calculator's statistical capabilities. Many calculators have these statistical functions built in to them. Needless to say, these functions can save you a lot of work and let you concentrate on the meaning of the numbers.

Key Words and Phrases

distribution	mean	median
mode	single-peaked	bimodal
symmetric	outliers	positively skewed
negatively skewed		

Key Concepts and Skills

- compute the mean, median, mode
- understand the effect of outliers on mean and median
- be aware of possible sources of confusion about "averages"
- determine the qualitative shape of a data set and identify whether it is skewed.

6 PROBABILITY: LIVING WITH THE ODDS

Overview

Probability is involved in nearly every decision we make. Often it is used on a subjective or intuitive level; occasionally we try to be more precise. As you will see in this chapter, probability is one of the older and most applicable branches of mathematics. Before we are done with the chapter, we will be able to apply probability to lotteries, gambling, life insurance problems and air traffic safety. It is a fascinating chapter, full of mathematics and real-life problems.

Unit 6A Fundamentals of Probability

Overview

We all have an intuitive sense of what a probability is and that sense is useful. A probability is a number between zero and one. If the probability of an event is zero, then it cannot happen; if the probability of an event is one, then it is certain to happen; and if the probability is somewhere is in between zero and one, it gives a measure of how likely the event is to happen.

We can approach probabilities in three different ways:

- a **theoretical probability** uses mathematical methods to find the probability of an event occurring.
- an **empirical probability** is determined with experiments or by data collection.
- a **subjective probability** is based on intuition.

This unit will focus on theoretical probability calculations; it will mention empirical probabilities, and we will leave subjective probabilities for another course!

If you wanted to find the probability that a fair dice will show a 5 when rolled once, you might reason that there are six possible outcomes of a single roll and a success (rolling a 5) is one of those six outcomes. So you might reason that the probability of rolling a 5 is 1/6. You would be correct! This thinking is the basic procedure for computing *a priori* probabilities:

Step 1: Count the total number of possible outcomes of an event.

Step 2: Count the number of outcomes that represent **success** — that is, the number of outcomes that represent the sought after result.

Step 3: Determine the probability of success by dividing the number of successes by the total number of possible outcomes:

$$\text{probability of success} = \frac{\text{number of outcomes that represent success}}{\text{total number of possible outcomes}}$$

This procedure can be used to compute the probability of many events, as you will see in the examples of the text. You will also see how the counting methods of the previous unit are used in steps 1 and 2 of this procedure.

Another important fact that will be used repeatedly is that if the probability of an event occurring is p, then the probability that it does not occur is $1 - p$. For example, the probability of rolling a fair die and getting a five is 1/6; therefore the probability of rolling *anything but* a five is 1 – 1/6 = 5/6.

Suppose you flip three coins and are interested in *all* of the outcomes. In this case, there are four possible outcomes: three heads, two heads and one tail, one head and two tails, and three tails. We will show how to find the probability of all of the outcomes. The result is called a **probability distribution,** which is just another way of saying the probabilities of *all* possible outcomes.

There are many situations in which it's impossible to determine probabilities *a priori*. In these situations, we can't use mathematical methods because we have only records or data to work with. For example, we often talk about a 100-year flood, which means that based on historical records, a flood of this magnitude occurs once every 100 years. So the probability of such a flood happening in any given year is 1/100. We discuss empirical probabilities because they are used so often in practice.

Finally, we take a few pages to clarify the confusion that arises over the use of the word *odds*. The odds *for* an event A are the probability of an event occurring divided by the probability of the event not occurring. For example, the odds of tossing a fair coin and getting a head are 1, often said *1 to 1*. Unfortunately, the term odds has an even slightly different meaning when used in gambling. There, it is the amount a winning bet will pay for every dollar that you bet. Needless to say, care must be used when dealing with odds (and gambling!).

Key Words and Phrases

event	outcome	theoretical probability
empirical probability	subjective probability	probability distribution
odds		

Key Concepts and Skills

- distinguish between theoretical probability, empirical probability, and subjective probability.
- use the three-step process to determine simple theoretical probabilities.
- find probability distributions for coin and dice experiments.
- determine empirical probabilities from data or historical records.
- find the probability of an event *not* occurring.
- find the odds of an event from its probability.

Important Review Boxes

- A Brief Review of the Multiplication Principle

Unit 6B Combining Probabilities

Overview

In the previous unit, we studied methods for finding the probability of individual events occurring. However, many interesting situations actually consist of multiple events. For example, what is the probability of tossing ten heads in a row with a fair coin? Or what is the probability of drawing a jack or a heart from a standard deck of cards? We will answer these and many other practical questions in this unit.

The unit presents five different techniques that involve multiple events. As always, it's important to know *how* to apply these techniques and *when* to apply them. The five situations covered by these methods are:

- independent AND events
- dependent AND events
- either/or mutually exclusive events (non-overlapping)

- either/or non-mutually exclusive events (overlapping)
- *at least once* events.

The term **AND event** refers to two or more events *all* happening. For example, for example what is the probability of rolling two fair dice and seeing a six on *both* dice (a six *and* a six)? Or what is the probability of drawing four cards from a standard deck and getting an ace each time? The important distinction is whether the events in question are **independent** or **dependent**. If one event does not affect the others (for example, in rolling two dice, the outcome of one die does not affect the outcome of other die), then we have a joint probability for independent events. The rule that applies for two independent events is

$$\text{P(A } \textit{and} \text{ B)} = \text{P(A)} \times \text{P(B)}.$$

This principle can be extended to any number of independent events. For example, the joint probability of A, B, and a third independent event C is

$$\text{P(A } \textit{and} \text{ B } \textit{and} \text{ C)} = \text{P(A)} \times \text{P(B)} \times \text{P(C)}.$$

If the events are not independent, then a bit more care is needed. For example, if you draw two cards from a deck, but don't replace the first card before drawing the second, then the outcome of the second card depends on the outcome of the first card. The rule that applies in this case for two dependent events is

$$\text{P(A } \textit{and} \text{ B)} = \text{P(A)} \times \text{P(B given A)}.$$

With **either/or events** we are interested in the probability of *either* event A *or* event B occurring. The important distinction is whether the events in question are mutually exclusive. If the occurrence of the first event prevents the occurrence of the second event, then the events are **mutually exclusive** or **non-overlapping**. For example, drawing a heart and drawing a diamond from a deck of cards are mutually exclusive events because if one event occurs (say, drawing a heart), then the other event (drawing a diamond) cannot occur. The rule that applies in this case with two events is

$$\text{P(A } \textit{or} \text{ B)} = \text{P(A)} + \text{P(B)}.$$

This principle can be extended to any number of mutually exclusive events. For example, the probability that either event A, event B, or event C occurs is

$$\text{P(A } \textit{or} \text{ B } \textit{or} \text{ C)} = \text{P(A)} + \text{P(B)} + \text{P(C)}.$$

If the events are **non-mutually exclusive** or overlapping (for example, going into a room of people and meeting a woman *or* a Democrat), then a modification must be used. The rule that applies in this case is

$$\text{P(A } \textit{or} \text{ B)} = \text{P(A)} + \text{P(B)} - \text{P(A } \textit{and} \text{ B)}.$$

In general, for AND probabilities we multiply probabilities and for either/or probabilities we multiply probabilities. However, modifications must be made in the cases of dependent or non-mutually exclusive events.

Finally, a very important situation in one in which we ask about the probability of an event happening *at least once*. For example, if you buy ten lottery tickets, what is the probability of *at least* one ticket being a winner? If you roll a die ten times, what is the probability of rolling *at least* one six? The rule that tells you the probability of an event A occurring at least once in *n* trials is

$$P(\text{A at least once in } n \text{ trials}) = 1 - P(\text{A does not occur in } n \text{ trials})$$
$$= 1 - P(\textit{not}\ \text{A})^n.$$

The unit has explanations of these rules and many examples to show how they are used. Be sure you refer to Table 8.2 for a summary of the rules. And be sure you also give yourself plenty of time to practice!

Key Words and Phrases

AND probability	independent events	dependent events
either/or probability	mutually exclusive	non-mutually exclusive
at least once rule		

Key Concepts and Skills

- distinguish between a joint probability and an either/or probability.
- distinguish between independent events and dependent events.
- distinguish between mutually exclusive events and non-mutually exclusive events.
- identify *at least once* situations.
- know when and how to apply the five probability rules given in the unit.

Unit 6C The Law of Averages

Overview

This unit is designed to strengthen your intuition about probabilities and to point out some common misconceptions about probabilities. It also deals with the important concept of expected value, which is important to understand if you gamble or buy life insurance.

We start by discussing the **law of averages**. We know that the probability of tossing a head with a fair coin is 1/2. This does not mean that if you toss a coin twice that you will always get one head and one tail. It does not mean that if you toss a coin ten times that you will always get five heads and five tails. What we *can* say is that the more often you toss the coin, the closer you should expect the fraction of heads and tails to approach 50%. This is an example of the law of averages.

The important concept in this chapter is **expected value**. Consider a situation in which there are several different outcomes. Each outcome has a known probability and a known cost or benefit. For example, a lottery may have a \$10 prize, a \$100 prize, a \$1000, and a grand prize of \$1,000,000. These are the four outcomes and each outcome has a certain probability. If you play this lottery many times, how much do you expect to win or lose in the long run? The answer is given by the expected value.

There is a general rule for computing expected values. In a situation with just two outcomes, the expected value is given by

$$\text{expected value} = \begin{pmatrix}\text{value of}\\ \text{event 1}\end{pmatrix} \times \begin{pmatrix}\text{probability of}\\ \text{event 1}\end{pmatrix} + \begin{pmatrix}\text{value of}\\ \text{event 2}\end{pmatrix} \times \begin{pmatrix}\text{probability of}\\ \text{event 2}\end{pmatrix}.$$

The unit presents several other applications of expected value as it arises in life insurance policies, lotteries, and the so-called house edge in casino gambling. There are some important practical lessons to be earned here.

The law of averages is related to the outlook called the **gambler's fallacy**. Gamblers often feel that if they get behind or start losing money, then their luck will change and their chances will improve. The fact is that probabilities do not change just because someone is losing. Once behind, it is likely that you will stay behind or get even further behind.

Key Words and Phrases

law of averages — expected value — gambler's fallacy — house edge

Key Concepts and Skills

- explain the law of averages.
- explain the gambler's fallacy.
- compute the expected value in situations with two or more outcomes.
- understand the house edge and compute it in specific situations.

Unit 6D Counting and Probability

Overview

We all know how to count *individual* objects or people, but there are many other types of counting problems that arise when we want to count *groups* of objects or people. That is the subject of this chapter. There are four basic counting methods presented in this chapter:

- selections from two or more groups (multiplication principle)
- arrangements with repetition
- permutations
- combinations.

In this unit, we will learn not only how to apply these methods, but equally important, how to determine *when* to use each method.

Selection from two or more groups occurs when we have several different groups of objects and we need select one item from each group. For example, in a restaurant, you might have four choices for an appetizer (first group), six choices for a main course (second group), and five choices for a dessert (third choice). How many different meals could you select?

Problems of this sort can be solved using a **table**, a **tree**, or (the easiest of all) the **multiplication principle** (see Review Box in Unit 7A). You will see that the total number arrangements of items from two or more groups is

(number of items in first group) × (number of items in second group) × (number of items in third group) × ···· × (number of items in last group).

In the above example, there would $4 \times 6 \times 5 = 120$ different possible meals that you could order.

Arrangements with repetition occur when you select items from a single group and items may be used more than once. Often the items in problems of this kind are letters or numerals. For example, how many different three-digit area codes can be formed from the numerals 0 through 9? There are 10 ways to choose the first numeral and, since repetition is allowed, there are 10 ways to choose the second numeral, and 10 ways to choose the third numeral. This amounts to a total of $10 \times 10 \times 10 = 1000$ different three-digit area codes. The general rule is that

> if we make r selections (for example, the three digits of the area code) from a group of n items (for example, the numerals 0–9), n^r different arrangements are possible.

Permutations also involve selecting items from a single group with one important difference — repetition is not allowed. Furthermore, in counting permutations, the order of the arrangements matters; that is, we count ABCD and DCBA as two different arrangements. To summarize, permutations require

- selection from a single group,
- repetition is not allowed, and
- order matters.

In the unit, we work slowly through several examples towards the general formula for counting permutations. We need a bit of mathematical notation along the way, so we introduce the **factorial** function.

$$n! = n \times (n-1) \times (n-2) \times \ldots \times 2 \times 1.$$

Your temptation might be to resist learning and using factorials, but they will make your life much easier! Just a little practice is all it takes.

With factorials in hand, we can write a general formula for counting permutations.

If we make r selections from a group of n items, the number of possible permutations is

$$_nP_r = \frac{n!}{(n-r)!},$$

where $_nP_r$ is read as "the number of permutations of n items taken r at a time."

Combinations are much like permutations with one difference — order does not matter; that is, we count ABCD and DCBA as the *same* arrangement. The requirements for combinations are

- selection from a single group,
- repetition is not allowed, and
- order does *not* matter.

The number of combinations of n objects taken r at a time is the same as the number of permutations, except that we have to correct for the over-counting that the permutation formula does (since permutations care about order and combinations don't). This subtle point is explained in the text. The result is the combinations formula.

If we make r selections from a group of n items, the number of possible *combinations*, in which order does not matter, is

$$_nC_r = \frac{_nP_r}{r!} = \frac{n!}{(n-r)! \times r!}$$

where $_nC_r$ is read as "the number of combinations of n items taken r at a time."

The key to mastering these methods is knowing not only *how* to use a particular method, but *when* to use it. There are plenty of practice problems at the end of the unit; you should work as many as possible. Another hint is to learn the capabilities of your calculator. You will need a calculator that at least does exponentiation (n^r) Many calculators compute factorials directly with a special factorial key. Some calculators have keys for permutations and combination. You will save yourself considerable work if you let your calculator do as much work as possible.

The unit concludes with a brief look at the subject of **coincidence**. Should you really be surprised when you win a raffle drawing at a basketball game? Should you really be surprised that after an evening of playing cards you are dealt a 13-card hand with eight hearts? Many so-called coincidences are bound to happen to someone, but we are always surprised when they happen to us.

We discuss the famous birthday problem and show that if there are 25 people in a room, then there is better than a 50% chance that two people have the same birthday. We also explore the phenomenon of **streaks**, particularly in sports, when a player repeats a certain success for many consecutive games. Should you really be surprised?

Key Words and Phrases

selection from two or more groups	multiplication principle	arrangements with repetition
permutation	combination	coincidence

Key Concepts and Skills

- describe the four different counting methods discussed in the unit.
- determine which of the four methods applies to a given counting problem.
- use factorials confidently.
- apply each of the four methods on appropriate problems.
- understand why not all coincidences should be surprising.

---

7 EXPONENTIAL ASTONISHMENT

Overview

If you had to learn just one lesson from a quantitative reasoning course, it might well be the difference between exponential growth and linear growth. Exponential growth and decay impacts our everyday lives in many ways, but most people are not aware of its presence or its power. In this chapter you will learn how everything from bank accounts to populations to radioactive waste are governed by exponential models; and you will learn how to construct exponential models to describe many everyday phenomena. In practical terms, this is an immensely important chapter.

Unit 7A Growth: Linear vs. Exponential

Overview

The difference between linear and exponential growth is stated on the first page of the chapter.

- *Linear growth* occurs when a quantity grows by the same *absolute* amount in each unit of time.
- *Exponential growth* occurs when a quantity grows by the same *relative* amount — that is, by the same *percentage* — in each unit of time.

Notice how these facts are related to the ideas of absolute and relative change studied in Unit 3A. It would be a good idea to contemplate these two statements and try to understand what they really mean. Hopefully this unit will also help!

This unit introduces the ideas surrounding exponential growth and decay in a very basic and accessible way. We use three parables (*From Hero to Headless*, *The Magic Penny*, and *Bacteria in a Bottle*) that illustrate quite dramatically the power of exponential growth. The goal is to develop some intuition about exponential growth and understand doubling processes.

The lessons of the unit are summarized in the highlight box at the end of the unit:

- Exponential growth is characterized by repeated doublings. With each doubling the amount of increase is approximately equal to the *sum* of all preceding doublings.

- Exponential growth cannot continue indefinitely. After only a relatively small number of doublings, exponentially growing quantities reach impossible proportions.

Key Words and Phrases

linear growth   exponential growth   doublings

Key Concepts and Skills

- explain the difference between linear and exponential growth.
- identify whether a given growth pattern is linear or exponential.
- understand the implications of a doubling process.

Unit 7B Doubling Time and Half-Life

Overview

Exponential growth is characterized by a constant **doubling time**. If a quantity (for example, a population or a bank account) is growing exponentially, then it doubles its size during a fixed period of time, and it continues to double its size over that same period of time forever. For example, if a tumor, growing exponentially, has a doubling time of two months, then it doubles its size during the first two months, and doubles its size again during the next two months, and continues to double in size *every* two months. Knowing the doubling time essentially defines the growth pattern for all times.

We denote the doubling time T_{double}. If we know T_{double} for a particular quantity that grows exponentially, then over a period of t time units, the quantity will increase by a factor of

$$2^{t/T_{double}}.$$

If we know the doubling time and the initial value of a particular quantity that grows exponentially, then we can find its value at all later times. The new values at later times are given by

$$\text{new value} = \text{initial value} \times 2^{t/T_{double}}.$$

If we know that a quantity grows with a constant percentage growth rate (for example, 5% per year), then we know it grow exponentially and that it has a constant doubling time. This suggests that there should be a connection between the percentage growth rate, which we call P, and the doubling time. We use a specific example of an exponentially growing population to present a widely used formula that relates percentage growth rate to doubling time. It is called the **Approximate Doubling Time Formula** or the **Rule of 70**. It says that

$$T_{double} \approx \frac{70}{P}.$$

This formula is an approximation and works best when the percentage growth rate is small (say, less that 10%). For example, if a bank account grows at 4% per year, it will double it value in approximately 70/4 = 17.5 years.

Everything we learned about exponential growth has a parallel with exponential decay. For example, if a quantity decays exponentially at a rate of 5% per month, it *decreases* by 5% during the first month, and by 5% during the second month, and continues to decrease by 5% every month. A quantity that decays exponentially has a constant **half-life** – the period of time over which it decreases its size by 50% or one-half.

We denote the doubling time T_{half}. If we know T_{half} for a particular quantity that decays exponentially, then over a period of t time units, the quantity will increase by a factor of

$$\left(\frac{1}{2}\right)^{t/T_{\text{half}}}$$

If we know the half-life and the initial value of a particular quantity that grows exponentially, then we can find its value at all later times. The new values at later times are given by

$$\text{new value} = \text{initial value} \times \left(\frac{1}{2}\right)^{t/T_{\text{half}}}.$$

The **Approximate Half-Life Formula** also applies to exponentially decaying quantities. If a quantity decreases by $P\%$ per unit time, the then half-life is given by

$$T_{1/2} \approx \frac{70}{P}.$$

As with the Approximate Doubling Time Formula, this half-life formula is approximate and works best when P is small (say, less than 10%).

You might wonder if there are exact formulae for finding the doubling time or half-life from the percentage growth or decay rates. For those who are curious, the unit closes with the *exact* doubling time and half-life formulae. These formulae are a bit more complicated and require the use of logarithms. They are exact for all percentage growth and decay rates, not just small ones.

Key Words and Phrases

doubling time	percentage growth rate	approximate doubling time formula
Rule of 70	half-life	percentage decay rate
approximate half-life formula	exact doubling time formula	exact half-life formula

Key Concepts and Skills

- identify the percentage growth rate from the description of an exponential growth process.
- find the doubling time from percentage growth rate.
- find the percentage growth rate from doubling time.
- determine the value of exponentially growing quantity given the doubling time and initial value.
- identify the percentage decay rate from the description of an exponential decay process.
- find the half-life from percentage decay rate.
- find the percentage decay rate from half-life.
- determine the value of exponentially decay quantity given the half-life and initial value.

Unit 7C Exponential Modeling

Overview

Having learned about the fundamentals of exponential growth and decay in Chapter 8, we can put these ideas to work to create models. Just as we used linear functions to model real world situations, we will now use exponential functions to model situations in which exponential growth or decay occur. We begin by introducing a general **exponential function**

$$Q = Q_0 \times (1 + r)^t,$$

where t represents time. This law require an initial value Q_0 and a growth or decay rate r. Notice that if r is positive the Q grows in time and if r is negative, then Q decays in time. Also important is that the units used for r and t must be the same (for example, if t has units of *days*, then r has units of 1/*days*). Once a specific exponential function is found, it can be used to predict the value of Q at all future times.

Be sure to study the highlight box entitled *Forms of the Exponential Function*. It shows that there are really several forms for the exponential function depending on whether you are given a growth rate, a doubling time, or a half-life.

With these two laws at hand, the rest of the unit is devoted to various applications. We look at how population growth, economic inflation, oil consumption, pollution, and drugs in the blood can all be modeled using these laws. Of particular importance is the technique called **radioactive dating**, which also relies on the exponential decay law. If these examples don't convince you of the widespread presence of exponential growth and decay, you will find even more applications in the problems!

Key Words and Phrases

exponential function   fractional growth rate   radioactive dating

Key Concepts and Skills

- given either a growth rate or a doubling time, use the appropriate form of the exponential growth law to model an exponentially growing quantity.
- given either a growth rate or a doubling time, use the appropriate form of the exponential growth law to model an exponentially decaying quantity.
- be familiar with forms of the exponential function that use the doubling time or half-life.
- understand radioactive dating and know how to determine the age of a material that contains a radioactive element.

Important Review Boxes

- A Brief Review of Algebra with Logarithms

---

8 MATHEMATICS AND THE ARTS

Overview

In this chapter we explore a very different application of mathematics, namely music and the fine arts. As you will see, the connections between mathematics and the arts go back to antiquity. However, with the development of digital music (CD players) and fractal art, the connections are also quite modern. We devote one unit to music, one unit to classical painting, and one unit to proportion as it appears in art and nature. This chapter should provide you with a new perspective and change of pace.

Unit 8A Mathematics and Music

Overview

The ties between mathematics and music go back to the ancient Greeks in about 500 B.C. The followers of Pythagoras discovered some of the basic laws that underlie our understanding of music today. They realized that the **pitch** of a musical note created by a plucked string is determined by the **frequency** of the string — how many times the string vibrates each second. They also discovered that if the length of a string is halved, the frequency doubles, and the pitch goes up by an **octave**. With these few facts we can explain a lot.

In a standard scale that you might play on a piano or a guitar, one octave consists of 12 tones or **half steps** (for example, the white and black keys between middle C and the next higher C). Here is the basic question we address: If we know the frequency of the first note of the scale, can we find the frequency of all 12 notes of the scale?

We show that to move up the scale a half step, we must *multiply* the frequency of the current tone by a fixed number. We give a brief argument showing that the magic number that generates the entire scale by multiplication is $f = 1.055946$ ... or the twelfth root of 2. It turns out the notes of the scale follow an exponential growth law, as discussed in Chapter 7.

The ancient Greeks understood that when two notes have a pleasing sound when played together (**consonant tones**), then the ratio of their frequencies must be the ratio of two small numbers, such as 3/2 or 4/3. We investigate how these ratios of small numbers compare to the exact frequencies generated by the magic number f.

The unit closes with a few observations about the modern connections between music and mathematics: digital music, compact disks, and synthesizers.

Key Words and Phrases

sound wave	pitch	frequency
cycles per second	fundamental frequency	overtones
octave	scale	half-step
consonant tones	analog	digital

Key Concepts and Skills

- understand the relation between frequency and pitch.
- determine the frequency of notes separated by an octave.
- determine the frequency of notes on a 12-tone scale, given the frequency of the first note.
- understand the Greek's explanation for consonant tones in terms of ratios of small integers.
- understand difference between analog and digital music.

Unit 8B Perspective and Symmetry

Overview

It wasn't until the Renaissance (14th and 15th century) that painters attempted to draw three-dimensional objects realistically on a flat two-dimensional canvas. It took many years for these painters to perfect this technique, but in the end, they made a science of **perspective** drawing. In this unit we trace the development of perspective drawing and explore some of its mathematical necessities.

The key concept in perspective drawing is the **principal vanishing point**. It can be summarized as follows: If you are an artist standing behind a canvas, painting a real scene, then all lines that are parallel in the real scene and perpendicular to the canvas, must meet in a single point in the painting — this is the principal vanishing point. All other sets of parallel lines in the real scene (that are *not* perpendicular to the canvas) meet in their own vanishing points. All of the vanishing points (principal and otherwise) lie along a single line called the **horizon line**.

Another fundamental property of paintings and other objects of art is **symmetry**. Symmetry can mean many different things, but it often refers to a sense of balance. We define three different kinds of symmetry:

- **reflection symmetry**: an object can be "flipped" across a particular line and it remains unchanged (for example, the letter H).
- **rotation symmetry**: an object can be rotated through a particular angle and it remains unchanged (for example, the letter O).
- **translation symmetry**: an object or a pattern can be shifted, say to the right or the left, and it remains unchanged (for example, the pattern ....XXXXXX.... extended in both directions).

We investigate these symmetries in both geometrical objects and in actual paintings.

Symmetry arises in beautiful ways in **tilings** — patterns in which one or a few simple objects are used repeated to fill a region of the plane. We give several examples of how triangles and quadrilaterals can be translated and reflected to produce wonderful patterns. And you can try some tilings in the problems, too!

Key Words and Phrases

vanishing point	principal vanishing point	horizon line
symmetry	reflection symmetry	rotation symmetry
translation symmetry	tiling	

Key Concepts and Skills

- understand the role of the principal vanishing point in perspective drawing.
- draw simple objects in perspective using vanishing points.
- identify symmetries in simple objects.
- draw simple objects with given symmetries.
- create tilings from triangles or quadrilaterals using translations and reflections.

Unit 8C Proportion and the Golden Ratio

Overview

In this unit, we explore another fundamental aspect of art; that is proportion. As you will see, issues of proportion arise not only in the art created by humans, but in natural forms as well. The subject has a lot to do with aesthetics, our innate sense of what is beautiful. And one of earliest statements about proportion and beauty goes back to the ancient Greeks who introduced the golden ratio.

The first instance of the golden ratio arises in dividing a line segment. What division of a line segment has the most visual appeal and balance? More specifically, suppose that the line segment has a length of L + 1 units and we want to divide it into two pieces of length L and 1. How should we choose L? The Greeks answered that the best division is the one that makes

$$\frac{\text{length of long piece}}{\text{length of short piece}} = \frac{\text{length of entire piece}}{\text{length of long piece}},$$

that is, the ratio of the length of the long piece to the length of the short piece is the same as the ratio of the length of the long piece to the length of the whole segment. We can also write this as

$$\frac{L}{1} = \frac{L+1}{L}.$$

We show that the value of L that makes this happen is the special (irrational) number

$$\phi = \frac{1+\sqrt{5}}{2} = 1.61803\ldots$$

This number is called the **golden ratio** or **golden section**.

From the golden ratio, we can define a **golden rectangle**. Any rectangle with sides in the ration of ϕ is called a golden rectangle. Both the golden ratio and the golden rectangle arise throughout the history of art. Great architectural works (for example, the Greek Parthenon) have dimensions close to those of the golden rectangle (although some claim that these examples are coincidences). Many common objects such as post cards and cereal boxes also have dimensions of golden rectangles. We also look at some examples of how the golden rectangle appears in nature.

There is one final connection that is too intriguing to ignore. We introduce the **Fibonacci sequence**, which was first used as a population model in the 13th century. Each term of the sequence is formed by adding the two previous terms:

$$1, 1, 2, 3, 5, 8, 13, 21, 34, \ldots..$$

We show how this sequence is related, perhaps unexpectedly, to the golden ratio. And the circle is closed by observing that the Fibonacci sequence also appears in the artwork of nature. This is a unit with several seemingly unrelated ideas that eventually become linked in a beautiful way.

Key Words and Phrases

golden ratio	golden rectangle	logarithmic spiral
symmetry	Fibonacci sequence	

Key Concepts and Skills

- understand the golden ratio as a proportion and divide a line segment according to the golden ratio.
- construct and identify golden rectangles.
- generate the Fibonacci sequence and find the ratio of successive terms.
- understand the connection between the golden ratio and the Fibonacci sequence.

---

9 MATHEMATICS AND POLITICS

Overview

This chapter explores the surprisingly crucial role that mathematics plays in political matters. The first unit deals with elections. We will se that in elections with more than two candidates, there are several ways to determine a winner, and they may not agree. In fact, it can be shown that there is no single voting method that meets certain fairness conditions. The second unit is devoted to another political process, namely apportionment. How do we determine the number of representatives that each state sends to Congress? Again, there are several methods that can be used, and they do not always agree!

Unit 9A Voting: Does the Majority Always Rule?

Overview

In this and the following unit we investigate mathematical problems associated with voting. This may sound like an unusual application of mathematics, but voting problems have been studied for several centuries and it has long been known that curious things can happen in voting systems. We will study such curious things in these two units.

The discussion begins with elections between two candidates. Throughout these units a candidate can be interpreted as a choice between two or more alternatives. For example, a candidate may be a person running for office or a brand of bagels in a taste test. With only two candidates, the rules are straightforward: the **majority** rules. This means that the candidate with the most votes (which must be more than 50% of the vote) wins the election.

However, even with majority rule, there are some interesting situations that can arise. We first look at presidential elections in which the winner is chosen, not by the **popular vote**, but by the **electoral vote**. Historically, there have been U.S. presidential elections in which a candidate won the popular vote, but lost the election.

We then look at variations on majority rule that often involve **super majorities**. For example, many votes in the U.S. government require more than a 50% majority: it takes a 2/3 super majority in both houses of Congress, followed by a 3/4 super majority vote of the states to amend the U.S. Constitution. More than a 50% majority of a jury is required to reach a verdict in a criminal trial.

Things get interesting when we turn to elections with three or more candidates. Often such elections are based on a **preference schedule** in which each voter ranks the candidates in order of preference. For example, the following preference schedule shows the outcome of an election among five candidates that we call A, B, C, D, and E.

First	A	B	C	D	E	E
Second	D	E	B	C	B	C
Third	E	D	E	E	D	D
Fourth	C	C	D	B	C	B
Fifth	B	A	A	A	A	A
	18	**12**	**10**	**9**	**4**	**2**

There was a total of 55 voters and 18 voters ranked A first, D second, E third, C fourth, and B last. The other columns are interpreted in a similar way. The question is: how do we determine a winner of the election?

The remainder of the unit presents five methods to determine a winner to an election with a preference schedule.

- **plurality**: the candidate with the most votes wins (candidate A would win the above election).
- **top two runoff**: the top two candidates have a runoff in which the votes of the losing candidates are redistributed to the top two candidates (candidate B would win the above election).
- **sequential runoff**: the candidates with the fewest first place votes are successively eliminated one at a time, votes are redistributed, and runoff elections are held at each stage (candidate C would win the above election).
- **point system** (or **Borda count**): with five candidates, five points are awarded for each first place vote, four points are awarded for each second place vote, and so on. The candidate with the most points wins (candidate D would win the above election).
- **pairwise comparison**: the winner between each pair of candidates is determined and the candidate with the most pairwise wins is the winner of the election (candidate E would win the above election).

You can probably already see the dilemma. We have proposed five reasonable methods for finding a winner and they all give different results! The unit closes inconclusively with this question unanswered. The next unit takes up the issue of fairness in voting systems and attempts to resolve the question.

Key Words and Phrases

majority rule	popular vote	electoral vote
super majority	preference schedule	plurality
top-two runoff	sequential runoff	point system
Borda count	pairwise comparisons	

Key Concepts and Skills

- understand the concept of majority rule and apply it in two -candidate elections.
- know the difference between popular vote and electoral vote.
- use super majority rules to determine the outcome of votes.
- create and interpret preference schedules.
- apply the methods of plurality, top-two runoff, sequential runoff, point system, and pairwise comparisons to determine the outcome of elections with preference schedules.

Unit 9B Apportionment: The House of Representatives and Beyond

Overview

The U.S. Constitution stipulates that each state should have representation in the House of Representatives proportional to its population. But it doesn't specify exactly how the number of representatives should be determined. The process of assigning representatives according to population, which has applications beyond the House of Representatives, is called **apportionment**.

Apportionment begins by computing a **standard divisor**, which is the average number of people in the population per representative. For example, if the total population is 100,000 and there are 100 representatives, then the standard divisor is 1000 people per representative. The next step is to compute the **standard quota** for each state, which is the number of representatives a state should have if fractional representatives were possible. For example, if a state has 5500 people, then with a standard divisor of 1000, that state should have 5500/1000 = 5.5 representatives. Now the dilemma of apportionment can be seen: it is not possible to have fractions of representatives!

The founding fathers proposed several methods to overcome this problem of fractional representatives. Here is a quick survey of the various methods that are considered in the text:

- **Hamilton's method**: First round the standard quota for each state *down* to form the **minimum quota** for that state; then give any remaining representatives to the states with the largest fraction of representatives.
- **Jefferson's method**: Choose a **modified divisor** such that when the new standard quotas are rounded *down* (to form a **modified quota**), all the representatives are used.
- **Webster's method**: Choose a **modified divisor** such that when the new standard quotas are rounded *according to the standard rounding rule* (to form a **modified quota**), all the representatives are used.
- **Hill-Huntington method**: Choose a **modified divisor** such that when the new standard quotas are rounded *according to a modified rounding rule* (to form a **modified quota**), all the representatives are used.

Notice that the methods appear quite similar and involve slightly different rules for rounding. However, as the examples in the text show, the methods can produce different results. Many examples of are provided to show how these methods are carried out.

Which method is best? Two hundred years of American history has proved that none of the methods is perfect. Each method can exhibit at least one of several paradoxes or can violate certain reasonable expectations. To address this question of fairness, the unit explores and gives examples of the following situations:

- the **Alabama paradox**: when the total number of representatives increases, but at least one state actually loses representatives.
- the **population paradox**: when the population increases and under reapportionment a slow-growing state gains representatives at the expense of a fast-growing state.
- the **new states paradox**: when additional representatives are added to accommodate a new state and the apportionment of existing states changes.
- the **quota criterion**: under any apportionment, the number of representatives for any state should be one of the whole numbers nearest its standard quota (the standard quote rounded either up or down).

The apportionment problem is a bit like the voting problem, in that a perfect apportionment method cannot be found. This is the conclusion of the **Balinsky and Young theorem**.

Key Words and Phrases

apportionment	standard divisor	standard quota
Hamilton's method	Alabama paradox	population paradox
new states paradox	Jefferson's method	modified divisor
modified quota	quota criterion	Webster's method
Hill-Huntington method	geometric mean	Balinsky and Young theorem

Key Concepts and Skills

- understand the problem and the dilemma of apportionment.
- apply the four methods of the text to apportionment problems.
- identify the deficiencies of the four methods in terms of paradoxes and the quota criterion.
- understand the significance of the Balinsky and Young theorem.

Unit 1A Recognizing Fallacies

Recognizing Fallacies.

1. Premise: Following Reagan's defense buildup, the Soviet Union began the process of democratization that ultimately led to its breakup. Conclusion: Reagan is responsible for the changes that led to the demise of the Soviet Union. Without historical substantiation, this is a false cause fallacy.

3. Premise: The mother has used drugs in the past. Conclusion: The decision on the custody case depends on the mother's present drug use. Limited choice.

5. Premise: Mom smoked when she was my age. Conclusion: I won't heed her advice to stop smoking. This is a circumstantial case of a personal attack. The fact that another person smokes is irrelevant to one's own personal decision to smoke.

7. Premise: No one has ever proved that telepathy doesn't exist. Conclusion: I believe in telepathy. This is an appeal to ignorance: an absence of proof for one conclusion (telepathy does not exist) does not prove the opposite conclusion (telepathy does exist).

9. Premise: Senator Smith is one of the biggest recipients of campaign contributions from the National Rifle Association. Conclusion: Senator Smith's bill cannot help the cause of gun control. This is a case of personal attack: the fact that the Senator gets campaign funds from the NRA does not necessarily mean that he or she will only introduce bills that are to the NRA's liking. (One could also argue that it's a kind of "guilt by association," implying that on the one hand, there exist politicians who have received funds from the NRA, some of whom have gone on to support the NRA's position, and on the other, that Sen Smith is one of these politicians. In this sense it's an implicit example of hasty generalization.)

11. Premises: The percentage of the population over 18 that smokes has decreased from 40% to about 20%. The percentage of overweight people has increased from 25% to 35%. Conclusion: Quitting smoking leads to overeating. This is false cause: just because there are now more overweight people and fewer smokers does not prove that quitting smoking causes overeating.

13. Premise: My little boy loves dolls and my little girl loves trucks. Conclusion: There's no truth to the claim that boys are more interested in mechanical toys while girls prefer maternal toys. This is an appeal to ignorance and hasty generalization.

15. Premise: Bush favors repealing the estate tax, which falls most heavily on the rich. Conclusion: Bush favors the rich; vote for Gore. This is a straw man argument.

Unit 1B Propositions and Truth Values

Is it a Proposition?

1. This statement is a proposition. It has a subject and a predicate, and makes a claim that is capable of being true or false.

3. This statement is a proposition. It has a subject and a predicate, and makes a claim that is capable of being true or false.

5. This question is not a proposition, it makes no claim.

Negation.

7. Negation: London is not the capital of England. The original statement is true; its negation is false.

9. Negation: Caesar was not a Roman. The original proposition is true; its negation is false.

11. Negation: some cowboys do not wear boots. The original is false; its negation is true.

13. Negation: no dogs are obedient. The original is true; its negation is false.

Multiple Negations.

15. The statement means that the city council supported the police chief, since they did not approve a vote *against* the police chief.

17. The statement means that Congress upheld the veto (i.e., opposed the bill), since it voted against a proposal that would have *reversed* the veto.

19. The statement means that the anti-discrimination policy is supported by the Constitution, since going against the policy is *contrary* to the spirit of the Constitution.

And Statements.

21. "Quebec is the capital of Canada" is false. "Moscow is the capital of Russia" is true. The conjunction is false.

23. "Ben is married" and "Ben is a bachelor" are the propositions, whose truth values cannot be assessed unless we know something about Ben. However, they cannot both be true, and so the conjunction is false.

25. "Grass is green" is true, as is "Skies are blue." The conjunction is true.

Interpreting *Or*.

27. This is the exclusive *Or*, you cannot chose both.

29. This could be either *Or*: exclusive if you only intend to make one trip, inclusive if you'd consider both (which you'll certainly be able to afford!).

31. This is the exclusive *Or*, if you assume that roads cannot be made of both asphalt and concrete. It's inclusive otherwise.

Or Statements.

33. "Seattle is the capital of the United States" is false, and "Paris is the capital of France" is true. The disjunction is true.

35. "Bill Clinton was a U.S. President" is true, and "Martin Luther King was a U.S. President" is false. The disjunction is true.

37. "$8 \times 4 = 32$ is true. "$7 \times 3 = 22$ is false. The disjunction is true.

Key Word Searches.

39. Search on "earthquakes" and "California."

41. Search on "Martin Luther King" and "assassination."

If... Then Statements.

43. The antecedent is "San Francisco is in California," which is true. The consequent is "San Francisco is in the Pacific Time Zone," which is also true. The implication is true.

45. The antecedent is "Paris is the capital of China," which is false. The consequent is

"London is the capital of England," which is true. The implication is true.

47. The antecedent is "Sparrows can fly," which is true. The consequent is "Sparrows are birds," which is also true. The implication is true.

49. The antecedent is "$2 + 2 = 5$," which is false. The consequent is "Primates are not mammals," which is also false. The implication is true.

Rephrasing Conditionals. In each of the following, p and q are (respectively) the antecedent and consequent in the given "If p then q" statements.

51. "If a person is a resident of Miami, then that person is a resident of the United States." This implication is true, since all residents of Miami (Florida) are automatically residents of the U.S.

53. "If you are a musician, then you play the saxophone." This proposition is false: not all musicians play the saxophone.

55. "If you have prostate cancer, then you are a male." This proposition is true, since only males have a prostate gland.

Converse, Inverse and Contrapositive.

57. The converse is "If Marco lives in the United States, then he lives in Chicago," which is false. The inverse is "If Marco does not live in Chicago, then he does not live in the United States," which, being logically equivalent to the converse, has the same meaning, and is also false. The contrapositive is "If Marco does not live in the United States, then he does not live in Chicago," which is true, being logically equivalent to the original proposition.

59. The converse is "If I get wet, then I went swimming," which is false. The inverse is "If I don't go swimming, then I won't get wet," which, being logically equivalent to the converse, has the same meaning, and is also false. The contrapositive is "If I don't get wet, then I didn't go swimming," which is true, being logically equivalent to the original proposition.

61. The converse is "If there is gas in the tank, then the engine is running," which is false. The inverse is "If the engine is not running, then there is no gas in the tank," which, being logically equivalent to the converse, has the same meaning, and is also false. The contrapositive is "If there is no gas in the tank, then the engine is not running," which is true, being logically equivalent to the original proposition.

Famous Quotes.

63. There are two conditional statements here. The first is "If a young man has not wept, then he is a savage." The second is "If an old man will not laugh, then he is a fool."

65. "If a person has no vices, then that person has few virtues."

Writing Conditional Propositions. Many solutions are possible here, we just suggest one example for each question.

67. If Sue lives in Cleveland, then she lives in Ohio.

69. If Sue lives in Ohio, then she lives in Cleveland.

71. Library Search.

a. Under a title search for "Finnegan," the book should be on the list since the title is *Finnegan's Wake.*

b. A search for "Finnegan" and "Joyce" should produce a list with *Finnegan' Wake* and any other works by or about Joyce with "Finnegan" in the title.

c. A search for "Joyce" or "Yeats" would provide a list of all works by anybody with the names Joyce or Yeats; the list would include *Finnegan's Wake.*

d. Searching for "Joyce" and "James" would give a list with all works by James Joyce and any other author with both James and Joyce in the name; so *Finnegan's Wake* would be on the list.

e. Searching for "Joyce" and "Yeats" would not produce *Finnegan's Wake* because both Joyce and Yeats must appear in the title/author string.

f. This time one would get a list of all works by Joyce or Yeats with "Finnegan" in the title; this list would include *Finnegan's Wake.*

73. Artificial Intelligence. If a computer exhibits human-level intelligence, then it must have the requisite processing power.

Truth Tables.

75. Comparing the two truth tables below, we see that "not (p or q)" is logically equivalent to "(not p) and (not q)."

p	q	p or q	not (p or q)
T	T	T	F
T	F	T	F
F	T	T	F
F	F	F	T

p	q	not p	not q	(not p) and (not q)
T	T	F	F	F
T	F	F	T	F
F	T	T	F	F
F	F	T	T	T

This has the same meaning as "neither," and together with the logical equivalence in Exercice 74, is known as De Morgan's Law.

77. Comparing the first truth table in Exercise 75 above with the truth table below, we see that "not (p or q)" is not logically equivalent to "not p or not q."

p	q	not p	not q	not p or not q
T	T	F	F	F
T	F	F	T	T
F	T	T	F	T
F	F	T	T	T

79. Comparing the two truth tables below, we see that "(p or q) and r" is logically equivalent to "(p and r) or (q and r)."

p	q	r	p or q	(p or q) and r
T	T	T	T	T
T	T	F	T	F
T	F	T	T	T
T	F	F	T	F
F	T	T	T	T
F	T	F	T	F
F	F	T	F	F
F	F	F	F	F

p	q	r	p and r	q and r	(p and r) or (q and r)
T	T	T	T	T	T
T	T	F	F	F	F
T	F	T	T	F	T
T	F	F	F	F	F
F	T	T	F	T	T
F	T	F	F	F	F
F	F	T	F	F	F
F	F	F	F	F	F

Unit 1C Sets and Venn Diagrams

Classifying Numbers.

1. a. $2.3 = \frac{23}{10}$ is a rational number.
b. $-\frac{2}{3} = \frac{-2}{3}$ is a rational number.
c. 3 is a natural number.
d. -5 is an integer.
e. $100.1 = \frac{1001}{10}$ is a rational number.
f. $-6.1 = \frac{-6.1}{10}$ is a rational number.

3. Set Notation.

a. {January, February, March, April, ..., November, December}.
b. {1, 2, 3, 4, ..., 98, 99, 100}.
c. {Washington, Adams, Jefferson}.
d. {Nebraska, Nevada, New Hampshire, New Jersey, New Mexico, New York, North Carolina, North Dakota}.

Venn Diagrams for Two Sets.

5. The sets *doctors* and *women* are overlapping sets, since it is possible for a person to be in one, both, or neither of these sets.

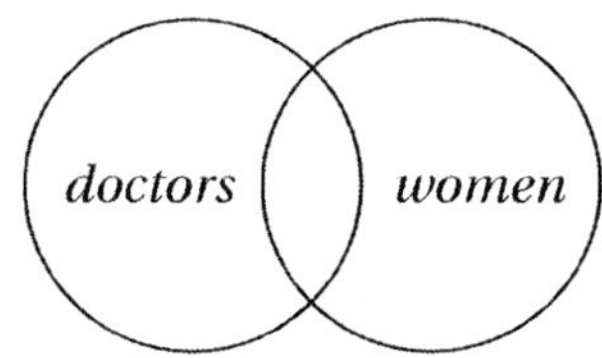

7. The sets *negative integers* and *natural numbers* are disjoint sets, since no number can be both negative and positive.

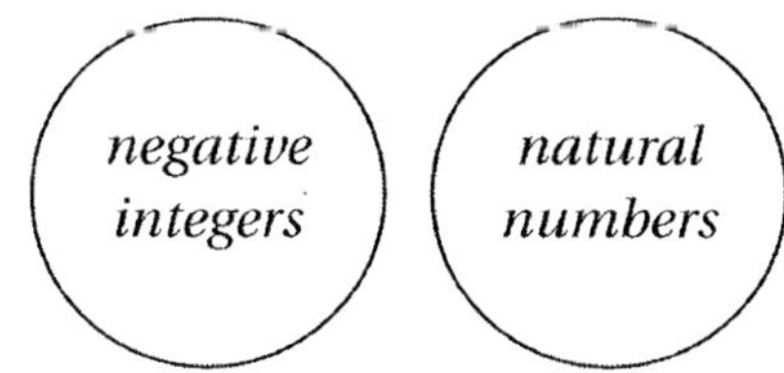

9. The set of *words* contains the set of *verbs*, since each verb is a word.

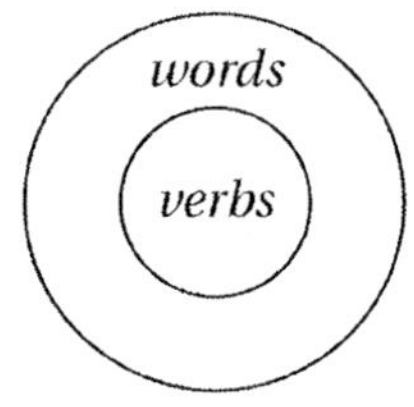

11. The set of *painters* is contained in the set of *artists*, since each painter is an artist.

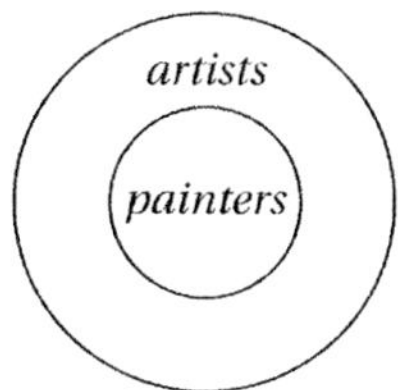

Categorical Propositions.

13. This is in standard form, with subject "bachelors" and predicate "men."

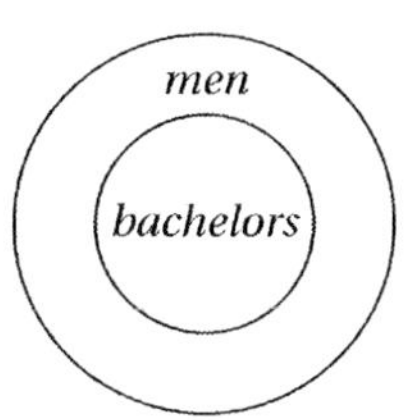

15. This is in standard form, with subject "U.S. presidents" and predicate "men."

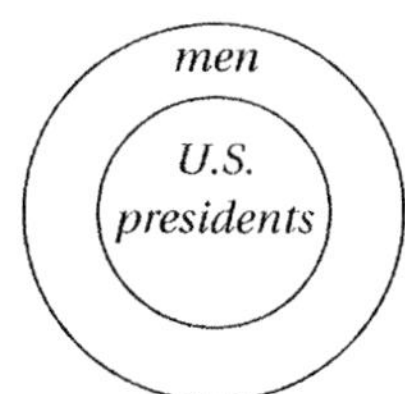

17. Standard form: "No fish are flying animals," with subject "fish" and predicate "flying animals."

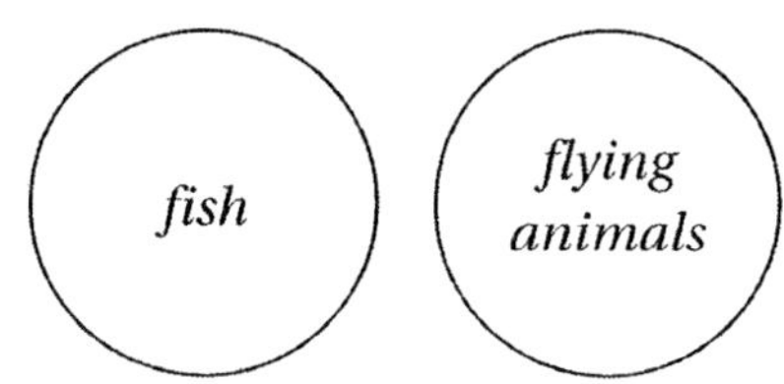

19. Standard form: "All nurses are people

who know CPR," with subject "nurses" and predicate "people who know CPR."

Categorical Negations.

21. This is in standard form, with subject "U.S. Senators" and predicate "republicans." The negation is "No U.S. Senators are Republicans." The original is true; its negation is false.

23. The standard form is "All doctors are left-handed people," with subject "doctors" and predicate "left-handed people." The negation is "Some doctors are not left-handed." The original is false; its negation is true.

25. The standard form is "Some people are sociable people," with subject "people" and predicate "sociable people." The negation is "No people are sociable people." The original is true; its negation is false.

Venn Diagrams for Three Sets.

27. Assuming that that no pilots are dentists, we get:

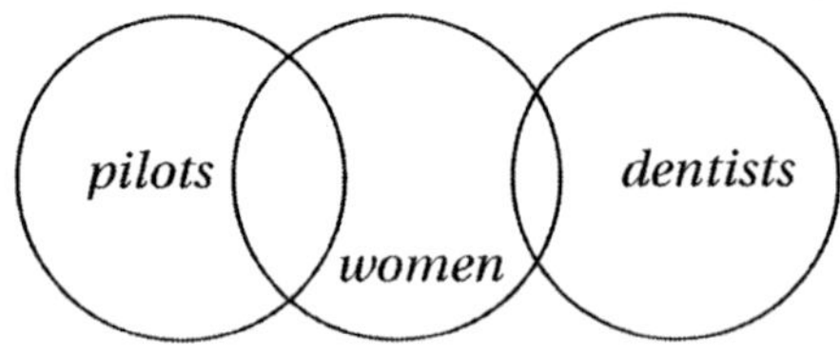

29. Assuming no songs are poetry and some poems and songs are unpublished, we get:

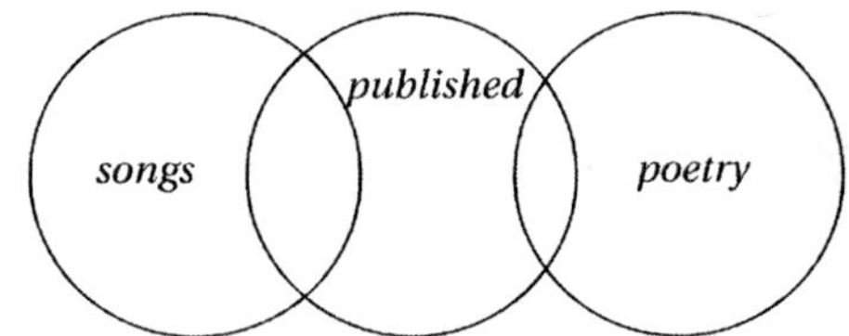

31.

hats
shirts
clothes

A Venn Diagram with Numbers.

33. a. There are $6 + 7 + 6 + 9 = 28$ people.

b. There are $6 + 9 = 15$ women.

d. There are $7 + 6 = 13$ men.

d. There are $7 + 6 = 13$ Republicans.

Venn Diagrams for Sets.

35.

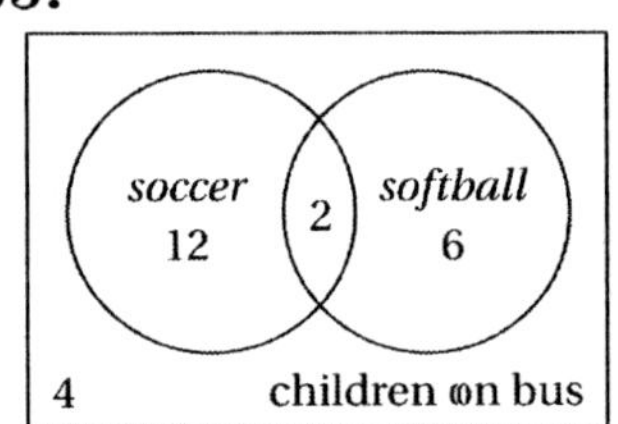

37.

drug
3
12
improvement
6
9
patients in study

Venn Diagrams Answers.

39.

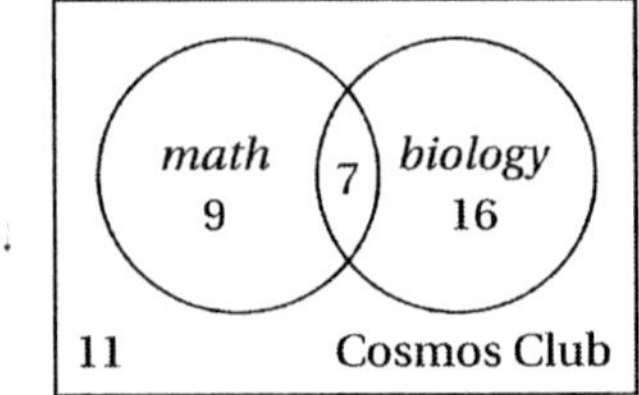

Since 16 students are taking at least math, and 9 are taking math only, then $16 - 9 = 7$ are taking math and biology. Since 23 are taking at least biology, then $23 - 7 = 16$ are taking biology only. Hence, $9 + 7 + 16 = 32$ are taking math or biology or both, and $43 - 32 = 11$ are taking neither.

41. Majors and Gender.

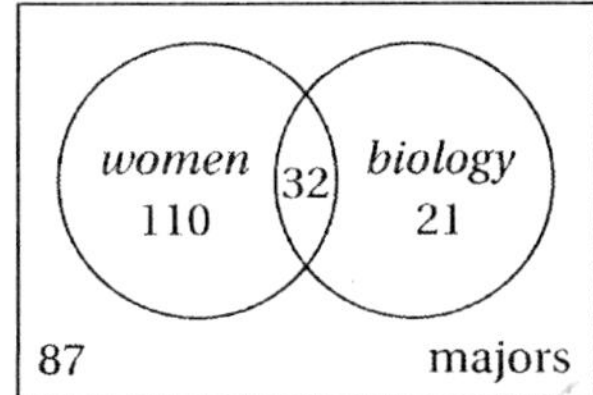

43. Coffee and Gall Stones.

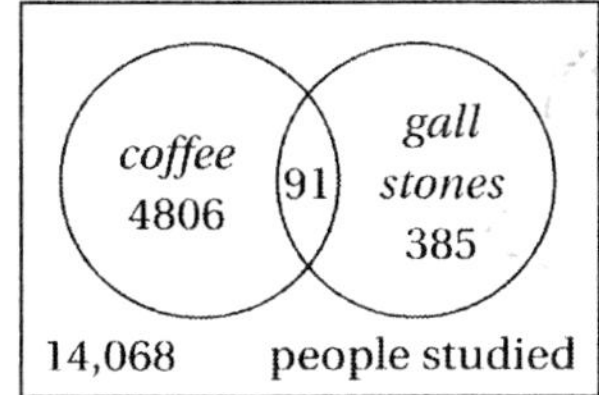

45. Interpreting Sales.

a.

b. $10 + 21 = 31$ people chose AC and 4WD: the 10 buyers who went for all three options and 21 who opted for AC and 4WD only.

c. $45 + 16 = 61$ people chose AC but not 4WD: the 45 who went for AC only and the 16 who opted for AC and CP only.

d. Summing over all of the non-overlapping option categories, we find that $45 + 21 + 8 + 16 + 10 + 8 + 12 = 120$ people chose (at least one of) AC or 4WD or CP.

e. $21 + 16 + 8 = 45$ people chose exactly two options: the 21 who chose AC and 4WD only, the 16 who chose AC and CP only, and the 8 who chose CP and 4WD only.

More Than Three Sets.

47.

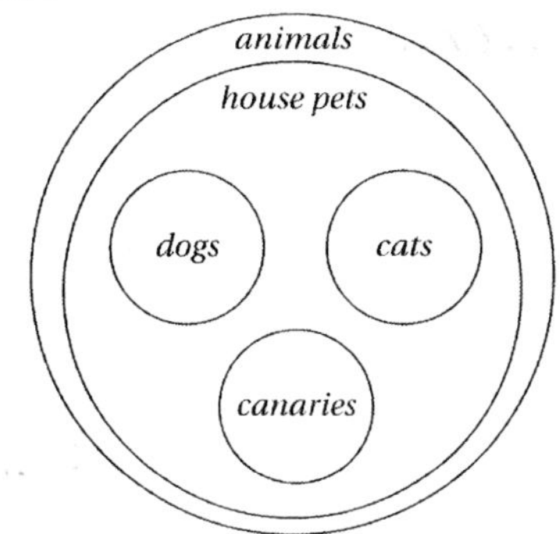

49.

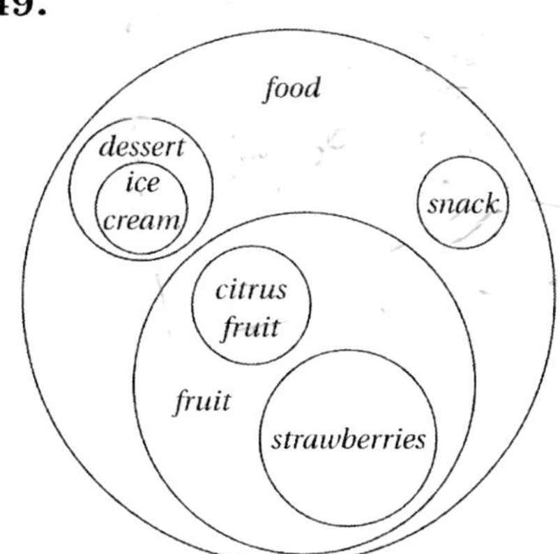

Organizing Propositions.

51.

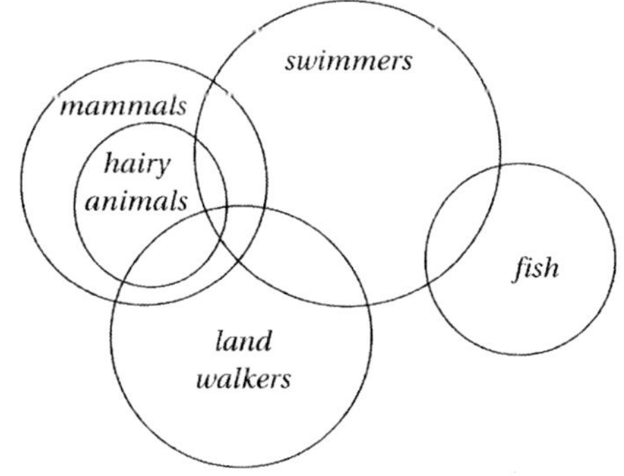

There could not be hairy fish. There could be hairy animals that swim. There could be mammals that walk on land. There could be hairy animals that walk on land.

53.

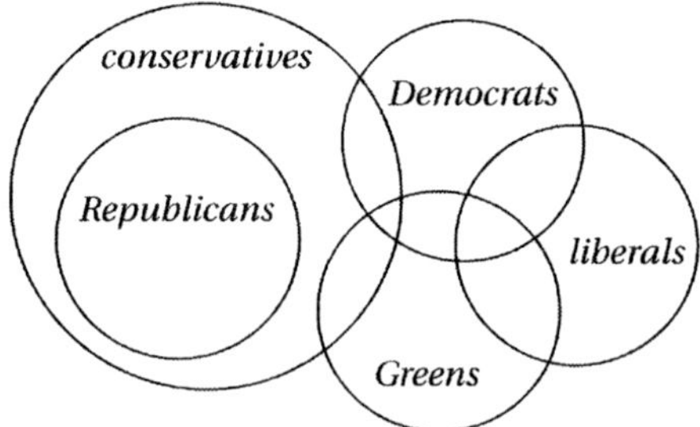

There could be conservative Democrats. There could be liberal Green Party members. There could not be liberal Republicans.

55. Is It Possible? Each shopper can be classified in one of $2 \times 2 \times 2 = 8$ possible ways, in terms of whether they approved or disapproved (two choices) of each of the three flavors. With two circles we only get $2 \times 2 = 4$ regions in our Venn diagram, which is not enough for the purposes at hand. It can be done, however, with three circles.

Unit 1D Critical Thinking in Everyday Life

Ambiguity in the News.

1. The quote is ambiguous because it is unclear if there were 350 people on each jet, or on both jets combined. If this could be clarified, the ambiguity would be removed.

3. The quote is ambiguous because it is unclear if the $83 billion refers to the gross domestic product or to 6% of the gross domestic product. If this could be clarified, the ambiguity would be removed.

5. Reading a Ballot Initiative.

a. The state constitution is less strict than the U.S. Constitution.

b. A yes vote would tighten (strengthen) laws against obscenity.

c. False.

7. Choosing Airline Tickets. Consider Plan A. If you go it costs $1600, and if you cancel it costs $400. With Plan B, if you go it costs $2900, and if you cancel it costs nothing. If you go, Plan B costs $1300 more. If you cancel, Plan A costs $400 more.

9. IRS Guidelines.

a. No, Trip need not file a federal tax return because both his earned and unearned incomes were well below the thresholds mentioned in conditions (i) and (ii) of the IRS guidelines, and he does not meet condition (iii) either.

b. No, Sally need not file a federal tax return because both her earned and unearned incomes were well below the thresholds mentioned in conditions (i) and (ii) of the IRS guidelines, and she does not meet condition (iii) either.

c. Yes, Monica must file a federal tax return because while her earned and unearned incomes were below the thresholds mentioned in conditions (i) and (ii) of the IRS guidelines, her total income of $\$3900 + \$300 = \$4200$ exceeds her earned income plus $250 (which is $\$3900 + \$250 = \$4150$ here), so that she meets condition (iii).

d. No, Horace need not file a federal tax return. His earned and unearned incomes were below the thresholds mentioned in conditions (i) and (ii) of the IRS guidelines, and his total income of $4000 + $200 = $4200 does not exceed his earned income plus $250 (which is $4000+$250 = $4250 here), so that he does not meet condition (iii) either.

e. Yes, Delila must file a federal tax return because while her earned and unearned incomes were below the thresholds mentioned in conditions (i) and (ii) of the IRS guidelines, her total income of $4200 + $200 = $4400 exceeds $4050 + $250 = $4300 here, so that she meets condition (iii).

11. What Is a Farmer?

a. Solutions will vary.

b. Jack does not appear to qualify because he works 150 hours over a period of less than five weeks; the regulation may be stated incorrectly.

c. Jill does appear to qualify because she works 150 hours over a period of more than five weeks; the regulation may be stated incorrectly.

d. No, Ivan must also satisfy either (c) or (d) under provision (1).

e. Yes, Joan satisfies three of the conditions under provision (1).

13. Comparing Candidates.

a. They could be consistent; Alice does not specify the time period for her 253 cases.

b. They could be consistent if Alice prosecuted 127 cases more than five years ago, most of which resulted in conviction, and 126 cases in the last five years, only four of which resulted in a conviction (as Zack claimed). This seems an unlikely explanation.

c. Solutions will vary.

15. A Financial Decision. If the estimate of the computer's resale value is accurate, the purchase option costs $1005, while the lease option costs $1050. All facts are "hard" except for the resale value.

Decision Making.

17. The package is the better choice.

19. If you go, plan (2) costs $250 more. If you cancel, plan (1) costs $75 more.

21. If you plan to fly 10 times (or a multiple of 10 times), then Plan A ($3150) is better than Plan B ($3250).

Critical Thinking.

23. Solutions will vary.

25. Solutions will vary.

27. Solutions will vary.

29. Solutions will vary.

31. Solutions will vary.

Unit 2A The Problem-Solving Power of Units

Working with Fractions.

1. a. $\frac{4}{3} \times \frac{1}{2} = \frac{4 \times 1}{3 \times 2} = \frac{4}{6} = \frac{2}{3}$.
b. $\frac{4}{3} + \frac{1}{2} = \frac{8}{6} + \frac{3}{6} = \frac{8+3}{6} = \frac{11}{6}$.
c. $\frac{4}{3} \div \frac{1}{2} = \frac{4}{3} \times \frac{2}{1} = \frac{4 \times 2}{3 \times 1} = \frac{8}{3}$.
d. $\frac{4}{3} - \frac{1}{2} = \frac{8}{6} - \frac{3}{6} = \frac{8-3}{6} = \frac{5}{6}$.
e. $\frac{7}{20} + \frac{3}{5} = \frac{35}{100} + \frac{60}{100} = \frac{35+60}{100} = \frac{95}{100} = \frac{19}{20}$.
f. $\frac{12}{13} - \frac{1}{4} = \frac{48}{52} - \frac{13}{52} = \frac{48-13}{52} = \frac{35}{52}$.
g. $\frac{7}{17} \times \frac{2}{7} = \frac{7 \times 2}{17 \times 7} = \frac{2}{17}$.
h. $\frac{1}{3} + \frac{1}{5} = \frac{5}{15} + \frac{3}{15} = \frac{5+3}{15} = \frac{8}{15}$.

3. a. $0.3 = \frac{3}{10}$.
b. $0.124 = \frac{124}{1000} = \frac{31}{250}$.
c. $0.78 = \frac{78}{100} = \frac{39}{50}$.
d. $0.005 = \frac{5}{1000} = \frac{1}{200}$.
e. $1.84 = \frac{184}{100} = \frac{46}{25}$.
f. $3.009 = \frac{3009}{1000}$.
g. $0.0001 = \frac{1}{10000}$.
h. $0.1001 = \frac{1001}{10000}$.

5. a. $\frac{3}{2} = 1.5$.
b. $\frac{4}{5} = 0.8$.
c. $\frac{1}{6} = 0.167$.
d. $\frac{1}{7} = 0.143$.
e. $\frac{2}{9} = 0.222$.
f. $\frac{7}{11} = 0.636$.
g. $\frac{88}{91} = 0.967$.
h. $\frac{122}{30} = 4.067$.

7. Identifying Units.
a. The price of apples will be in \$ per pound.
b. The speed will be in km/s, namely in kilometers per second.
c. The installation cost will be in $\$/\text{ft}^2$, namely in dollars per square feet.
d. The flow rate will be in ft^3/s, namely in cubic feet per second.
e. The atmospheric pressure will be in lb/in^2, namely in pounds per square inch.

9. Area and Volume Calculations.
a. The area of the pool's surface is $10 \text{ m} \times 5 \text{ m} = 50 \text{ m}^2$. The pool's volume is $10 \text{ m} \times 5 \text{ m} \times 3 \text{ m} = 150 \text{ m}^3$.
b. The area of one of its two largest sides is $22 \text{ in} \times 15 \text{ in} = 330 \text{ in}^2$. The package's volume is $22 \text{ in} \times 15 \text{ in} \times 12 \text{ in} = 3960 \text{ in}^3$.
c. The skyscraper's volume is $1000 \text{ ft} \times 25,000 \text{ ft}^2 = 25,000,000 \text{ ft}^3$.

11. Unit Calculations.
a. To convert 18 yards to feet we use an appropriate conversion factor:

$$18 \text{ yd} \times \frac{3 \text{ ft}}{1 \text{ yd}} = 54 \text{ yd}.$$

b. To convert 18 yards to feet we use two conversion factors:

$$18 \text{ yd} \times \frac{3 \text{ ft}}{1 \text{ yd}} \times \frac{12 \text{ in}}{1 \text{ ft}} = 648 \text{ in}.$$

c. To convert 142 ounces to pounds we use an appropriate conversion factor:

$$142 \text{ oz} \times \frac{1 \text{ lb}}{16 \text{ oz}} = 8.875 \text{ lb}.$$

d. To convert $\frac{1}{5}$ acre to square feet we use an appropriate conversion factor:

$$\frac{1}{5} \text{ acre} \times \frac{43,560 \text{ ft}^2}{1 \text{ acre}} = 8712 \text{ ft}^2.$$

e. To convert 6 gallons to ounces, we use a chain of conversion factors:

$$6 \text{ gal} \times \frac{4 \text{ qt}}{1 \text{ gal}} \times \frac{4 \text{ cup}}{1 \text{ qt}} \times \frac{8 \text{ oz}}{1 \text{ cup}} = 768 \text{ oz}.$$

13. Squared Units.

a. Since 1 foot equals 12 inches, 1 square foot equals 12^2 square inches. So $1 \text{ ft}^2 = 144 \text{ in}^2$. This can also be written as $\frac{144 \text{ in}^2}{1 \text{ ft}^2}$ or $\frac{1 \text{ ft}^2}{144 \text{ in}^2}$.

b. To convert 5 square feet to square inches, we use one of the conversion factors from (a):

$$5 \text{ ft}^2 \times \frac{144 \text{ in}^2}{1 \text{ ft}^2} = 720 \text{ in}^2.$$

c. The area is $100 \text{ yd} \times 60 \text{ yd} = 6000 \text{ yd}^2$. To convert this to square feet, we need to use the fact that 1 yard is 3 feet, and hence 1 square yard is 9 square feet. Hence,

$$6000 \text{ yd}^2 = 6000 \text{ yd}^2 \times \frac{9 \text{ ft}^2}{1 \text{ yd}^2} = 54,000 \text{ ft}^2.$$

15. Currency Conversions.

a. 1 British pound is worth about 1.5144 dollars, which is more than 1 dollar.

b. To convert \$100 to yen, note that

$$\$100 = \$100 \times \frac{107.53 \text{ yen}}{\$1} = 10,753 \text{ yen}.$$

c. Converting 2500 pesos to dollars yields

$$2500 \text{ pesos} = 2500 \text{ pesos} \times \frac{\$0.1073}{1 \text{ peso}} = \$268.25.$$

d. First we convert 5.50 euros to dollars:

$$5.50 \text{ euros} = 5.50 \text{ euros} \times \frac{\$0.9544}{1 \text{ euros}} = \$5.25.$$

Hence, gasoline in Germany sells for \$5.25 per gallon.

Working with Units.

17. If you buy 4.7 pounds of apples priced at \$1.29 per pound, you will pay:

$$4.7 \text{ lb} \times \frac{\$1.29}{1 \text{ lb}},$$

which is about \$6.06.

19. If you travel 14 miles in 15 minutes, then your speed in miles per hour is:

$$\frac{14 \text{ miles}}{15 \text{ min}} \times \frac{60 \text{ min}}{1 \text{ hr}} = 56 \frac{\text{miles}}{\text{hr}}.$$

21. If you paid \$200,000 for 40 acres of land, then you paid:

$$\frac{\$200,000}{40 \text{ acres}} = \frac{\$5000}{1 \text{ acre}}.$$

23. The cost of 10 square yards of cloth, priced at \$2 per square yard, is:

$$10 \text{ yd}^2 \times \frac{\$2}{1 \text{ yd}^2} = \$20.$$

25. The floor area is 1520 square feet, which we must first convert to square yards (see Exercise 13 above):

$$1520 \text{ ft}^2 \times \frac{1 \text{ yd}^2}{9 \text{ ft}^2} = 168.89 \text{ yd}^2.$$

Carpeting it will cost:

$$168.89 \text{ yd}^2 \times \frac{\$18}{\text{yd}^2} = \$3040.$$

27. If you travel 1200 miles in 20 hours, then your average speed for the trip is:

$$\frac{1200 \text{ miles}}{20 \text{ hr}} = 60 \frac{\text{miles}}{\text{hr}}.$$

29. If you sleep (on average) 7.5 hours each night, then in a year you sleep:

$$7.5 \frac{\text{hr}}{\text{night}} \times 365 \text{ nights},$$

which is 2737.5 hours.

What Went Wrong?

31. Exam Question. Sorry! You are wrong because your answer has units of 1/$ ("per dollars"). The correct answer is 0.11 lb $\times \frac{\$7.70}{1 \text{ lb}} = \$0.85$, i.e., 85 cents.

33. Exam Question. Sorry! You are wrong because your large bag price answer has units of pounds per dollar. The correct large bag price is $\frac{\$11}{5 \text{ lb}} = \frac{\$0.22}{\text{lb}}$, i.e., 22 cents per pound. Thus, the large bag is much cheaper *per pound* than the small bag, which costs 39 cents per pound.

35. Filling a Pool. Since 6 inches is 0.5 feet, the volume of water needed to fill the pool is 75 ft $\times$ 54 ft $\times$ 0.5 ft $= 2025 \text{ ft}^3$.

37. Full of Hot Air. If an average human heart breaths 6 times per minute, exhaling half a liter of hot air each time, then since $6 \times 0.5 = 3$, we see that 3 liters of hot air are exhaled per minute. This translates into

$$3 \frac{\text{L}}{\text{min}} = 3 \frac{\text{L}}{\text{min}} \times 60 \frac{\text{min}}{\text{hr}} \times 24 \frac{\text{hr}}{\text{day}},$$

i.e., 4320 liters per day.

39. Dog Years. a. 15 real years = 15 real years $\times \frac{7 \text{ dog years}}{1 \text{ real year}} = 105$ dog years.

b. The third year in the life of a human child starts when the child is two years old, and ends when the child is three years old. If dogs have a similar period, from their second to third (dog) birthdays, note that

$$2 \text{ dog years} = 2 \text{ dog years} \times \frac{1 \text{ real year}}{7 \text{ dog years}},$$

or 0.2857 real years, which comes out to 0.2857 years $\times \frac{52 \text{ weeks}}{\text{year}} = 15$ weeks, and

$$3 \text{ dog years} = 3 \text{ dog years} \times \frac{1 \text{ real year}}{7 \text{ dog years}},$$

or about 0.4286 real years, which is 0.4286 years $\times \frac{52 \text{ weeks}}{\text{year}} = 22$ weeks. Thus, "the terrible twos" for dogs is the period from about 15 to 22 weeks in their lives.

Unit 2B Standardized Units: More Problem-Solving Power

Powers of 10.

1. a. $10^6 \times 10^5 = 10^{6+5} = 10^{11} = 100,000,000,000$.

b. $10^4 \times 10^{-3} = 10^{4-3} = 10^1 = 10$.

c. $\frac{10^6}{10^5} = 10^{6-5} = 10^1 = 10$.

d. $\frac{10^4}{10^{-3}} = 10^{4-(-3)} = 10^7 = 10,000,000$.

e. $10^{-2} \times 10^{-4} = 10^{-2-4} = 10^{-6} = 0.000001$.

f. $\frac{10^{-7}}{10^{-9}} = 10^{-7-(-9)} = 10^2 = 100$.

g. $10^6 + 10^5 = 1,000,000 + 100,000 = 1,100,000$.

h. $10^7 - 10^2 = 10,000,000 - 100 = 9,999,900$.

USCS Units.

3. a. Solutions will vary of course, but for somebody who is 5 feet 6 inches, we get (5 ft $\times \frac{12 \text{ in}}{\text{ft}}$) + 6 in = (60 + 6) in = 66 inches.

b. Since 128 oz $\times \frac{1 \text{ lb}}{16 \text{ oz}} = 8$ lb, a gallon of water weighs about 8 pounds.

c. Since

$$\frac{1}{2} \text{ gal} \times \frac{8 \text{ pts}}{1 \text{ gal}} = 4 \text{ pt},$$

the jug holds 4 liquid pints. Equivalently, it holds

$$4 \text{ pt} \times \frac{28.88 \text{ in}^3}{1 \text{ pt}} \times \frac{1 \text{ dry pt}}{33.60 \text{ in}^3} = 3.4381 \text{ dry pts},$$

or about 3.44 dry pints.

d. 150 bushels is equivalent to

$$150 \text{ bushels} \times \frac{4 \text{ pecks}}{1 \text{ bushel}} \times \frac{8 \text{ dry qt}}{1 \text{ peck}} \times \frac{67.2 \text{ in}^3}{1 \text{ dry qt}},$$

or, about 322,560 cubic inches. Hence, 150 million bushels of wheat is equivalent to about 322,560 million cubic inches of wheat.

Since $1 \text{ ft}^3 = (12 \text{ in})^3 = 1728 \text{ in}^3$, then

$$322,560 \text{ in}^3 = 322,560 \text{ in}^3 \times \frac{1 \text{ ft}^3}{1728 \text{ in}^3},$$

which is about 187 cubic feet. Hence, 150 million bushels of wheat is about 187 million cubic feet.

5. Metric Prefixes. a. A millimeter is 1000 times smaller than a meter, since 1000 millimeters is 1 meter.

b. A gram is 1000 times smaller than a kilometer, since 1000 grams is 1 kilogram.

c. A square millimeter is 1,000,000,000,000 times smaller than a square kilometer. This is because 1000 millimeters is 1 meter and and 1000 meters is 1 kilometer, so that there are $1000^2 = 1,000,000$ millimeters in a kilometer, and hence $1,000,000^2 = 1,000,000,000,000$ square millimeters in a square kilometer.

7. USCS-Metric Conversions.

a. Converting yields $10 \text{ m} \times \frac{3.28 \text{ ft}}{1 \text{ m}}$, which is 32.80 ft.

b. Converting, we get

$$880 \text{ yd} \times \frac{0.9144 \text{ m}}{1 \text{ yd}} \times \frac{1 \text{ km}}{1000 \text{ m}},$$

which is about 0.8 kilometer.

c. Converting yields $20 \text{ gal} \times \frac{3.785 \text{ l}}{1 \text{ gal}}$, which is about 75.7 liters.

d. Converting, we get

$$5 \text{ mL} \times \frac{0.03381 \text{ fl oz}}{1 \text{ mL}} \times \frac{1.805 \text{ in}^3}{1 \text{ fl oz}},$$

which is about 0.3 cubic inches.

e. First note that as 1 foot is 0.3048 meters, then $1 \text{ ft}^2 = (0.3048 \text{ m})^2 = 0.0929 \text{ m}^2$, and so $1200 \text{ ft}^2 = 1200 \text{ ft}^2 \times \frac{0.0929 \text{ m}^2}{1 \text{ ft}^2}$, which is about 111.5 square meters.

9. Celcius-Fahrenheit Conversions.

a. To convert 45 degrees Fahrenheit to Celcius, subtract 32 to get 13, then divide by 1.8 to get 7.2 degrees Celcius.

b. To convert 20 degrees Celcius to Fahrenheit, multiply by 1.8 to get 36, then add 32 to get 68 degrees Fahrenheit.

c. To convert -15 degrees Celcius to Fahrenheit, multiply by 1.8 to get -27, then add 32 to get 5.0 degrees Fahrenheit.

d. To convert -30 degrees Celcius to Fahrenheit, multiply by 1.8 to get -54, then add 32 to get -22 degrees Fahrenheit.

e. To convert 70 degrees Fahrenheit to Celcius, subtract 32 to get 38, then divide by 1.8 to get 21. degrees Celcius.

11. Celcius-Kelvin Conversions.

a. To convert 50 degrees Kelvin to Celcius, subtract 273.15 to get -223.15 degrees Celcius.

b. To convert 240 degrees Kelvin to Celcius, subtract 273.15 to get -33.15 degrees Celcius.

c. To convert 10 degrees Celcius to Kelvin, add 273.15 to get 283.15 degrees Kelvin.

Sensible or Ridiculous?

13. This statement is sensible: liters is indeed a measure of liquid, and 2 liters of water is about 2 quarts, or 4 pints, which is a reasonable amount to drink in one day.

15. This statement is sensible: kilometers is a measure of distance, and 100 kilometers per hour is about 62 miles per hour—which is a reasonable speed at which to drive on an interstate highway.

17. This statement is ridiculous: while meters is a measure of height, 7 meters is over 20 feet—nobody can jump that high!

19. This statement is ridiculous: while milligrams is a measure of weight, 3 milligrams is "featherlight"—no book weighs that little.

21. The Metric Mile.

a. In order to compare the metric mile to the USCS mile we must find a common set of units. Note that 1 mile is 1.6093 kilometers, or 1609.3 meters, and $1500 - 1609.3 = -109.3$ meters. So, a metric mile is 109.3 meters shorter than a USCS mile, and since $-\frac{109.3}{1609.3} = -0.0679$, the metric mile is about 6.8% shorter than a USCS mile.

b. Solutions will vary.

23. Tallest Mountain? The total height of Mauna Kea from its ocean floor base to its peak is $13,796 + 18,200 = 31,996$ feet. Converting to miles and kilometers, in turn, we get $31,996 \text{ ft} \times \frac{1 \text{ mi}}{5280 \text{ ft}} = 6.06$ mi, and $6.06 \text{ mi} \times \frac{1.6093 \text{ km}}{1 \text{ mi}} = 9.75$ km.

The total height of Mauna Kea is thus greater than the above sea level height of Mount Everest (29,023 feet), but to assert that the Hawaiian mountain is taller is to ignore the fact that there is more to Mount Everest than meets the eye; a fair comparison should take into account its own rise from the ocean floor.

Gems and Gold.

25. A nugget that is 25% gold is 6-karat gold, since 100% is 24-karat gold.

27. A 2.5-carat diamond weighs $2.5 \times 0.2 = 0.5$ grams.

Electric Bills.

29. a. 1250 kilowatt-hours is 1250×3.6 million joules, which is 4500 million, or 4.5 billion, joules.

b. The average power used in watts is the total number of joules used per second. Since there are

$$60 \frac{\text{s}}{\text{min}} \times 60 \frac{\text{min}}{\text{hr}} \times 24 \frac{\text{hr}}{\text{day}} \times 30 \frac{\text{days}}{\text{month}},$$

i.e., 2,592,000 seconds in June, the average power used is:

$$\frac{4,500,000,000 \frac{\text{j}}{\text{month}}}{2,592,000 \frac{\text{s}}{\text{month}}},$$

which is about 1736 joules per second, i.e., about 1736 watts (1.736 kilowatts).

c. If each liter of burned oil releases 12 million joules of energy, then to produce 4500 million joules we'll need to burn

$$4500 \text{ million j} \times \frac{12 \text{ million j}}{1 \text{ L}},$$

which is 375 liters. This can be checked to be about 99 gallons.

31. Basketball Power. Burning 800 calories per hour can be converted to joules per second as follow:

$$\frac{800 \text{ Cal}}{1 \text{ hr}} \times \frac{4184 \text{ j}}{1 \text{ Cal}} \times \frac{1 \text{ hr}}{60 \text{ min}} \times \frac{1 \text{ min}}{60 \text{ s}},$$

i.e., about 930 joules per second, or 930 watts. This is enough to keep nine 100-watt bulbs shining.

33. Refrigerator Cost. A 350-watt refrigerator consumes energy at the rate of 350 joules per second. Since there are

$$\frac{60\text{ s}}{1\text{ min}} \times \frac{60\text{ min}}{1\text{ hr}} \times \frac{24\text{ hr}}{1\text{ day}} \times 365\text{ days},$$

i.e., 31,536,000 seconds in a year, the unit uses $31,536,000 \times 350 = 11,038$ million joules per year. Dividing by 3.6 million joules per kilowatt-hour yields 3066 kilowatt-hours. Each kilowatt-hour costs $0.08, so the total cost is $\$0.08 \times 3066 = \$245.28$.

35. Compact Flourescent Light Bulbs. Using the 25-watt bulb in place of the 100-watt one saves 75 watts, and since there are

$$\frac{60\text{ s}}{1\text{ min}} \times \frac{60\text{ min}}{1\text{ hr}} \times 10,000\text{ hr}$$

i.e., 36 million seconds in 10,000 hours, the bulb substitution saves 75×36 million, or 2700 million, joules. Dividing by 3.6 million joules we get 750 kilowatt-hours. Since each kilowatt-hour costs $0.08, the total saving over the lifespan of the fluorescent bulb is $\$0.08 \times 750 = \$60$.

37. Coal Power Plant. Since there are

$$\frac{60\text{ s}}{1\text{ min}} \times \frac{60\text{ min}}{1\text{ hr}} \times \frac{24\text{ hr}}{1\text{ day}} \times 30\text{ days}$$

i.e., 2.592 million seconds in a month, and the plant can generate 1 billion watts of power, i.e., 1 billion joules per second, in a month it will generate 1 billion j $\times$ 2.592 million, or 2592 million million (i.e., trillion), joules. Dividing by 3.6 million joules we get 720 million kilowatt-hours.

If each kilogram of coal yields about 450 kilowatt-hours of energy when burned, then to fuel this power plant each month will require: 720 million kilowatt-hours $\times \frac{1\text{ kg}}{450\text{ kilowatt-hours}}$, which comes out to 1.6 million kilograms of coal.

If a typical home used 1000 kilowatt-hours of energy per month, then this power plant can supply: 720 million kilowatt-hours $\times \frac{1\text{ home}}{1000\text{ kilowatt-hours}}$, which is 720,000 homes.

Solar Energy.

39. A 1-square-meter panel with 20% efficiency generates 20% of 1000, i.e., 200 watts of power. This 200 watts of power is 200 joules per second, and since there are $6 \times 60 \times 60 = 21,600$ seconds in 6 hours, we see that each day (which yields 6 hours of direct sunlight) leads to the production of $200 \times 21,600 = 4,320,000$ joules of energy.

The 200 watts of power which the panels can generate only applies to the 6 hours out of 24 which they are actually receiving direct sunlight. So, on average, their power is $\frac{6}{24} \times$ 200 watts, namely 50 watts.

41. Project: Personal Enery Audit Solutions will vary.

Unit 2C Problem-Solving Guidelines and Hints

Traffic Counters.

1. a. We seek all possible pairs (M, N) of M three-axle Trucks and N two-axle Cars which could result in the 35 axle counts registered by the traffic counter. (Algebraically, this means that $M \times 3 + N \times 2 = 35$.) As discussed in Example 1 in the text, one solution is $(M, N) = (11, 1)$, and another is $(M, N) = (9, 4)$. Just as there is no solution with $M = 12$, there is none with $M = 10$, as this would account for 30 axle counts, leaving 5 axle counts for two-axle cars, which is not possible. Similarly, there are no solutions for any even value of M. Since we found solutions for $M = 11, 9$, we try other odd values $M = 7, 5, 3, 1$. Each of these works, leading to four new solutions $N = 7, 10, 13, 16$, respectively. Overall, we get six possible combinations $(M, N) = (11, 1), (9, 4), (7, 7), (5, 10), (3, 13), (1, 16)$.

b. Here we seek all possible pairs (M, N) of M three-axle Trucks and N two-axle Cars which could result in 41 axle counts registered by the traffic counter. (This means that $M \times 3 + N \times 2 = 41$.) Let's try $M = 11$ again: this accounts for 33 axle counts, leaving 8 axle counts for two-axle cars, which means that $N = 4$. So $(M, N) = (11, 4)$ is one solution. In fact $M = 13$ also works: this accounts for 39 axle counts, leaving 2 axle counts for two-axle cars, which means that $N = 1$, so $(M, N) = (13, 1)$ is another solution. As before, there is no solution with $M = 12$, as this would account for 36 axle counts, leaving 5 axle counts for two-axle cars. In fact, there are no solutions for any even value of M. Let's try other odd values $M = 9, 7, 5, 3, 1$. Each of these works, leading to $N = 4, 7, 10, 13, 16$, respectively. We overlooked one: $M = 13$, which leads to $N = 1$. There are seven combinations $(M, N) = (13, 1), (11, 4), (9, 7), (7, 10), (5, 13), (3, 16), (1, 19)$, in total.

3. If Jill's time in the first race was 10 seconds, then her pace was 100 m $\div$ 10 s = 10 m/s, or 10 meters per second. Jack's pace would therefore be 95 $\div$ 10 s = 9.5 meters per second. For the second race, Jill's time would be 105 m $\div 10\frac{\text{m}}{\text{s}}$ = 10.5000 seconds, and Jack's would be be 100 m $\div 9.5\frac{\text{m}}{\text{s}}$ = 10.5263 seconds, so Jill would win by 0.0263 seconds. Note how the assumptions here amount to a doubling of the paces, and hence a halving of the times taken to run the race, and a halving of the winning margin.

5. The distance between the cars is shrinking at a rate of 180 kilometers per hour (the combined speed of the two cars), and hence the cars will meet after 150 km $\div$ $180\frac{\text{km}}{\text{hr}} = \frac{5}{6}$ hours. Meanwhile, the canary, flying at 120 kilometers per hour, will have flown $\frac{5}{6}$ hrs $\times$ 120 $\frac{\text{km}}{\text{hr}}$ = 100 kilometers.

7. Mixing Marbles. Pile 1 originally starts as (15B,0W), Pile 2 as (0B,15W), and three marbles are transferred from the first to the second pile. Thus, Pile 1 becomes (12B,0W) and Pile 2 (3B,15W). Next, three marbles are transferred back from Pile 2 to Pile 1. Four possibilities result, depending on how many of these three are black, as tabulated below.

Transfers from 2 to 1	New Pile 1	New Pile 2
3B	(15B,0W)	(0B,15W)
2B	(14B,1W)	(1B,14W)
1B	(13B,2W)	(2B,13W)
0B	(12B,3W)	(3B,12W)

Note how in each case the number of black marbles in Pile 2 equals the number of white marbles in Pile 1.

9. China's One-Child Policy. If 10,000 families have children according to the one-son policy, then 5000 families have a son as their first, and therefore only child. Of the other 5000 families who had a daughter as their first child, 2500 have a son as their second child, and no more children. Of the other 2500 families, who had a daughter as their second (as well as first) child, 1250 have a son as their third child, and no more children. Of the other 1250 families, who had a daughter as their third (as well as first and second) child, 625 have a son as their fourth child, and no more children. Of the other 625 families, who had a daughter as their fourth child, about 312 have a son as their fifth child, and no more children. Of the other 312 families, who had a daughter as their fifth child, about 156 have a son as their sixth child, and no more children. Of the other 156 families, who had a daughter as their sixth child, about 78 have a son as their seventh child, and 78 had a daughter, and so on.

In this scenario, keeping track up to the seventh child, we see that the total number of sons is $5000 + 2500 + 1250 + 625 + 312 + 156 + 78 = 9921$. The total number of girls is also $5000 + 2500 + 1250 + 625 + 312 + 156 + 78 = 9921$! So boy and girls are being born in equal numbers, and since there are $9921 + 9921 = 19,842$ total for 10,000 families, this yields an average of 1.9842 children per family. If we do the bookkeeping past the seventh child, assuming that families continue having children until a boy is born, the average gets closer and closer to 2.

Puzzle Problems.

11. If it takes 30 seconds to walk between the first and third floors of a building, then it takes 15 seconds to walk up one floor. Hence, to walk from the first to the sixth floor takes five times 15, i.e., 75 seconds.

13. If you draw one, two or three apples from the basket, they could all be of different types. However, once you draw a fourth you must have (at least) two of one kind, since there are only three kinds available.

15. If I am somebody's brother, yet have no brothers myself, then that somebody (the blind fiddler in this case) must be a woman.

17. By selecting a fruit from the box labeled *Apples and Oranges*—which being incorrectly labelled must contain *only* apples or oranges—you immediately learn how that box should be labelled. Let's suppose you selected an apple, then the box labelled *Oranges* must contain both apples and oranges, and the box labelled *Apples* must contain only oranges. On the other hand, if you selected an orange, then the box labelled *Oranges* must contain only apples and the box labelled *Apples* must contain both apples and oranges. Thus, with just one selection, you can determine the correct labelling for all three boxes. Note, however, that if you start by selecting from one of the other boxes you learn less and must make a second selection, so our solution is the best possible.

19. Clearly the woman must ferry the animals across one at a time, being careful to never leave unattended any creatures which can harm each other. Her first crossing must therefore me to take the goose to the other side. She returns alone, and picks up either of the other animals for a second crossing to the other side. However, she needs to return *with* the goose, which she exchanges for the one remaining creature for the third crossing to the other side. It is safe to leave the mouse and wolf together on the other side, and she goes back one last time to pick up the goose again. After her fourth crossing to the other side she has made seven crossings in total, counting the three return trips.

21. Since half of the rope length is 75 feet, and the midpoint of the rope is 25 feet from the ground and the poles are 100 feet high, the ropes must be vertical and the flagpoles right next to each other (no distance between them).

23. If you draw one or two socks from the drawer, they could all be different colors. However, once you draw a third sock you must have (at least) two of one color, since there are only two colors available.

25. Prisoner 1 cannot see red hats on both Prisoners 2 and 3; if he did, he would know he had a white hat (and would be freed). Therefore the hats on Prisoners 2 and 3 must be either both white, or one of each color, and Prisoner 2 knows this! Therefore if Prisoner 2 sees a red hat on Prisoner 3, Prisoner 2 knows he has a white hat (and would be freed). Since each of Prisoners 1 and 2 claim to be unable to determine the color of his own hat, Prisoner 3 can deduce that he has a white hat, and he is freed.

27. Solutions will vary.

29. Solutions will vary.

31. Solutions will vary.

33. Solutions will vary.

35. Solutions will vary.

Unit 3A Uses and Abuses of Percentages

Review of Fractions.

1. a. $1/4 = 0.25 = 25\%$.
b. $0.45 = 45\% = 9/20$.
c. $0.33333\ldots = 33.333\ldots\% = 1/3$.
d. $23\% = 0.23 = 23/100$.
e. $1/8 = 0.125 = 12.5\%$.
f. $1.34 = 134\% = 67/50$.
g. $0.65 = 65\% = 13/20$.
h. $99\% = 0.99 = 99/100$.

Review of Ratios.

3. a. The ratio of A to B is $\frac{A}{B} = \frac{10}{2} = 5$, the ratio of B to A is $\frac{B}{A} = \frac{2}{10} = 0.2$, and A is 500% of B.
b. The ratio of A to B is $\frac{A}{B} = \frac{150}{25} = 6$, the ratio of B to A is $\frac{B}{A} = \frac{25}{150} = 0.167$, and A is 600% of B.
c. The ratio of A to B is $\frac{A}{B} = \frac{11}{1} = 11$, the ratio of B to A is $\frac{B}{A} = \frac{1}{11} = 0.91$, and A is 1100% of B.
d. The ratio of A to B is $\frac{A}{B} = \frac{45}{550} = 0.0818$, the ratio of B to A is $\frac{B}{A} = \frac{550}{45} = 12.22$, and A is 8.18% of B.
e. The ratio of A to B is $\frac{A}{B} = \frac{75}{480} = 0.1563$, the ratio of B to A is $\frac{B}{A} = \frac{480}{75} = 6.4$, and A is 15.63% of B.
f. The ratio of A to B is $\frac{A}{B} = \frac{33}{123} = 0.2683$, the ratio of B to A is $\frac{B}{A} = \frac{123}{33} = 3.73$, and A is 26.83% of B.

Percentages as Fractions.

5. a. 25 is 22.12% of 113, because $\frac{25}{113} = 0.2212$.
b. 345 is 86.68% of 398, because $\frac{345}{398} = 0.8668$.
c. 1234 is 41.01% of 3009, because $\frac{1234}{3009} = 0.4101$.

Percentage Change.

7. The absolute change is $1509 - 2226 = -717$ newspapers, and the percentage change is $-\frac{717}{2226} \times 100\% = -32.21\%$.

9. The absolute change is $\$850 - \$160 = \$690$ billion, and the percentage change is $\frac{\$690}{\$160} \times 100\% = 431.25\%$.

11. The absolute change is $30\% - 12\% = 18\%$, which is an 18 percentage point rise, and the percentage change is $\frac{18\%}{12\%} \times 100\% = 150\%$.

13. The absolute change is $33004 - 28638 = 4366$ libraries, and the percentage change is $\frac{4366}{28638} \times 100\% = 15.25\%$.

Percentage Difference.

15. Taking the daily circulation figure for the NYT as the reference value, the absolute difference is $1.77 - 1.07 = 0.7$ million newspapers, and the percentage difference is $\frac{0.7}{1.07} \times 100\% = 65.42\%$.

17. Taking the U.S. tourist figure as the reference value, the absolute difference is $66 - 48 = 18$ million international arrivals, and the percentage difference is $\frac{18}{48} \times 100\% = 37.50\%$.

19. Taking the female open-heart surgery figure as the reference value, the absolute difference is $504000 - 209000 = 295000$ operations, and the percentage difference is $\frac{295000}{209000} \times 100\% = 141.15\%$.

Of vs. More Than.

21. Jack's weight is $\frac{100+40}{100} = 1.4$ times Jill's, so his weight is 140% of hers.

23. The population of Montana is $\frac{100-20}{100} = 0.80$ times the population of New Hampshire, so Montana's population is 80% of New Hampshire's.

25. The wholesale cost is (100-50)% = 50% of the retail cost, i.e., 0.5 times the retail cost. So the retail cost is $\frac{1}{0.50} = 2$ times the wholesale cost.

27. The retail cost is (100+40)% = 140% of the wholesale cost, i.e., 1.4 times the wholesale cost.

29. The absolute difference between the 1975 and 1997 percentages of high school seniors using alcohol is 68.2% – 52.7% = −15.5%, this is a 15.5 percentage point drop. In relative terms, the difference is $-\frac{15.5\%}{68.2\%} = -0.2273$, which reflects a 22.73% drop.

31. The absolute difference between the 1960s and 1990s five-year survival rate for Caucasians for all forms of cancer is 60% - 39% = 21%, so there was a 21 percentage point rise. In relative terms, the difference is $\frac{21\%}{39\%} = 0.5385$, which reflects a 53.85% rise.

33. Assume 30% of city employees in Carson City ride the bus to work. If "the percentage of city employees in Freetown who ride the bus to work is *10% higher* than in Carson City," then the percentage of city employees in Freetown who ride the bus to work is $\frac{100+10}{100} \times 30\% = 33\%$. On the other hand, if "the percentage of city employees in Freetown who ride the bus to work is *10 percentage points higher* than in Carson City," then the percentage of city employees in Freetown who ride the bus to work is 10% + 30% = 40%. In the first case, we are told how much higher the Freetown rate is as a percentage *of* the Carson City rate, so we add 10% *of* the lower rate to get the higher rate. In the second case, we are told how much higher than the Carson City rate the Freetown rate in absolute percentage points, which means that we add 10 percentage points *to* the lower rate to get the higher rate.

35. A bicycle with a $699 price tag will cost $\frac{100+7.6}{100} \times \$699 = \$752.12$ when subjected to a local sale tax rate of 7.6%.

37. A car with a $17,600 price tag will cost $\frac{100+6.5}{100} \times \$17600 = \$18,744$ when subjected to a local sale tax rate of 6.5%.

39. Leaving $22 for an $18.75 dinner bill means that your tip was $22 - $18.75 = $3.25 in absolute terms. In relative terms, you tipped at a rate of $\frac{\$3.25}{\$18.75} = 0.1733 = 17.33\%$.

41. If the original smoking rate R jumped 45 percent, to 18.3 percent, then R satisfies: $\frac{100+45}{100} \times R = 1.45 \times R = 18.3\%$. Hence, $R = \frac{18.3\%}{1.45} = 12.6\%$.

43. Since 75% of $100 billion is $75 billion, the value of U.S. trade with Mexico that is delivered by truck is almost $75 billion.

45. A 20% growth in a sales value of $3 billion will lead to $\frac{100+20}{100} \times \$3$ billion = $3.6 billion being spent on bagels next year.

47. A 39% growth in consumer debt of D leads to $\frac{100+39}{100} \times D = 1.39 \times D$. Since this is (approximately) $1 trillion, D is (approximately) $\frac{\$1\text{ trillion}}{1.39} = \$0.7194$ trillion.

Shifting Reference Value.

49. False. Her weight after losing 10% was 0.90 times her original weight, and following the subsequent 10% weight gain it went up to $(1.10)(0.90) = 0.99$ times—i.e., 99% *of*—her original weight, which is 1% less than her original weight.

51. False. Your final taxes are $(1.06)(1.05) = 1.1130$ times—or 111.30% *of*—your original salary, which is 11.3% more than your original taxes.

53. Not sensible. One price cannot be more than 100% less than another price.

55. Sensible; e.g., if the average house price increased from \$100,000 to \$550,000.

57. Sensible; this means that your computer speed is three and a half times mine.

59. No, you cannot simply average the averages for classes of different sizes.

61. True; blond haired women comprise $(0.10)(0.60) = 0.06 = 6\%$ of the class.

63. False, some of the hotels with restaurants may also have pools. (If all of them do, only 70% of the hotels in town have a restaurant and a pool.)

Unit 3B Putting Numbers in Perspective

Review of Scientific Notation.

1. a. $5 \times 10^6 = 5,000,000 = 5$ million.
b. $7 \times 10^9 = 7,000,000,000 = 7$ billion.
c. $2 \times 10^{-2} = 0.02 = 2$ hundredths.
d. $8 \times 10^{11} = 800,000,000,000 = 800$ billion.
e. $1 \times 10^{-7} = 0.0000001 = 1$ ten millionth.
f. $9 \times 10^{-4} = 0.0009 = 9$ ten thousandths.

3. a. $600 = 6.0 \times 10^2$.
b. $0.9 = 9 \times 10^{-1}$.
c. $50,000 = 5 \times 10^4$.
d. $0.003 = 3 \times 10^{-3}$.
e. $0.00005 = 5 \times 10^{-4}$.
f. $70,000,000,000 = 7.0 \times 10^{10}$.

5. a. $(3 \times 10^4) \times (8 \times 10^5) = (3 \times 8) \times (10^4 \times 10^5) = 24 \times 10^9 = 2.4 \times 10^{10}$.
b. $(6.3 \times 10^2) + (1.5 \times 10^1) = 630 + 15 = 645 = 6.5 \times 10^2$.
c. $(9 \times 10^3) \times (5 \times 10^{-7}) = (9 \times 5) \times (10^3 \times 10^{-7}) = 45 \times 10^{-4} = 4.5 \times 10^{-3}$.
d. $(4.4 \times 10^{99})/(2 \times 10^{11}) = (4.4/2) \times (10^{99}/10^{11}) = 2.2 \times 10^{88}$.

They Don't Look That Different!

7. a. 10^{35} is 10^9, or 1 billion, times larger than 10^{26}.
b. 10^{27} is 10^{10}, or 10 billion, times larger than 10^{17}.
c. 1 billion $= 1 \times 10^9$ is 10^3, or 1000, times larger than 1 million $= 10^6$.
d. 7 trillion $= 7 \times 10^{12}$ is 10^9, or 1 billion, times larger than 7 thousand $= 7 \times 10^3$.
e. 2×10^{-6} is 10^3, or 1 thousand, times larger than than 2×10^{-9}.
f. 6.1×10^{29} is 10^2, or 100, times larger than 6.1×10^{27}.

Using Scientific Notation.

9. Corporate profits in the United States in 1996 were $\$6.32 \times 10^{11}$.

11. The hard drive on my computer has a capacity of 5.4×10^9 bytes.

13. The area of the Earth's surface is 5.096×10^8 square kilometers.

Approximation with Scientific Notation.

15. a. No rounding or estimating (or calculator!) is needed in this problem. $20,000 \times$

$100 = (2\times10^4)(1\times10^2) = (2\times1)(10^4\times10^2) = 2\times10^6 = 2,000,000$.

b. We could approximate this quotient by $(9\times10^3)/(3\times10^1) = 9/3\times10^2 = 300$. A calculator yields 311; so our estimate was reasonably close.

c. Note that the answer here will be negative. We can approximate this product by $-12\times12\times10^3 = -144\times10^3 = -1.44\times10^5$. The actual value is about -1.49×10^5; our estimate was reasonably close.

d. Again, no rounding or estimating is needed. 250 million $\times 40 = (2.5\times10^8)(4\times10^1) = 2.5\times4\times10^9 = 10\times10^9 = 10^{10}$, or 10 billion.

e. We could approximate this product by $7\times300 = 2100$, which is not too far from the actual value of about 2169.

f. We could approximate this quotient by $(6.6\times10^6)/(3.3\times10^1) = 6.6/3.3\times10^5 = 2\times10^5 = 200,000$. A calculator yields about \$200,948, so our estimate was not too far off.

Perspective Through Estimating.

17. A 10-story building is between 100 and 120 feet high. A football field is 100 yards—or 300 feet—long, which is almost three times longer than the building is high.

19. It is not unusual for Americans to put 15,000 to 20,000 miles on a single car in a year. Assume an average American (excluding traveling business people) takes two airplane trips per year of 5000 miles. This amounts to 10,000 miles of flying. To the extent that these assumption are accurate, we can conclude that Americans drive more than they fly each year.

21. We will find the amount spent on movies by estimating the number of people who go to the movies, how often they go, and the amount spent on admission tickets. Let's assume that 100 million of the U.S. population of 280 million go to the movies. If these people go 10 times per year, on average, and pay \$7 for admission, on average, then the total spent per year will be

$$10^8 \text{ persons} \times \frac{10 \text{ movies}}{\text{person} \times \text{year}} \times \frac{\$7}{\text{movie}}$$

which works out at \$7 billion per year. Our ticket price estimate reflects big city prices, and is probably too high as a national average, especially if we take lower children's prices into account. With a \$5 ticket price average, and assuming that 150 million people attend movies an average of 20 times a year, our estimate would jump to \$15 billion per year. It is difficult to give a precise estimate with more accurate information concerning movie going habits and ticket prices.

23. If we assume that you take about 10 breaths per minute (i.e., one every six seconds), then we would take

$$\frac{10 \text{ breaths}}{\text{min}} \times \frac{60 \text{ min}}{\text{hour}} \times \frac{24 \text{ hour}}{\text{day}} \times \frac{7 \text{ day}}{\text{week}}$$

which is 100,800 breaths per week.

25. To estimate the number of words in the textbook, we could multiply the number of pages in the book times the average number of words per page. The average number of words per page can, in turn, be estimated by multiplying the average number of lines per page by the average number of words per line. Now estimate the number of pages, the average number of lines per page, and the average number of words per line. Suppose the book has 700 pages. A few sample pages suggest that the average number of lines per page is 40 and the average number of words per line is 15. Therefore, the total number of words per text is approximately

$$\frac{700 \text{ pages}}{\text{text}} \times \frac{40 \text{ lines}}{\text{page}} \times \frac{15 \text{ words}}{\text{line}}$$

which is 420,000 words per text.

The accuracy of our estimate is good if all three numbers needed for the estimate are accurate; these could be determined more accurately by sampling more pages and lines.

27. There are several ways to estimate here. Some people could estimate the amount a family spends on food more easily than the amount spent by an individual on food. Should we include meals eaten in restaurants? Some people rarely eat out, others do it daily. If an individual spends \$50 per week, then that individual would spend

$$\frac{\$50}{\text{week}} \times \frac{52 \text{ weeks}}{\text{year}} = \frac{\$2560}{\text{year}}.$$

This estimate is subject to error due to the variation in the amount spent per week.

Making Numbers Understandable.

29. A birth rate of approximately 4 million per year in the U.S. translates to

$$\frac{4 \times 10^6}{\text{year}} \times \frac{1 \text{ year}}{365 \text{ days}} \times \frac{1 \text{ day}}{24 \text{ hours}} \times \frac{1 \text{ hours}}{60 \text{ mins}},$$

or, approximately 7.6 births per minute.

31. A birth rate of approximately 132 million per year worldwide translates to

$$\frac{132 \times 10^6}{\text{year}} \times \frac{1 \text{ year}}{365 \text{ days}} \times \frac{1 \text{ day}}{24 \text{ hours}} \times \frac{1 \text{ hours}}{60 \text{ mins}},$$

or, approximately 251 births per minute.

33. Experiment with a stack of crisp new dollar bills; the higher the better. (It's easier to measure a stack of fifty bills than a stack of ten, and it will lead to more accurate results.) The thickness of a single bill can then be estimated by dividing the height measured by the number of bills involved. It's also a good idea to use metric units here, to avoid a string of messy conversions. Let's assume that you stack up fifty \$1 bills and it is about 0.5 cm high. Advertising spending of approximately \$200 billion then translates to a stack of dollar bills which is

$$220 \times 10^9 \text{ bills} \times \frac{0.5 \text{ cm}}{50 \text{ bills}} \times \frac{1 \text{ km}}{10^5 \text{ cm}},$$

or, approximately 22,000 kilometers high. This would hardly be feasible, as it is more than three times the radius of the earth!

35. An annual heart attack death rate of about 720,000 in the U.S. translates to

$$\frac{720,000 \text{ deaths}}{\text{year}} \times \frac{1 \text{ year}}{365 \text{ days}},$$

or, about 1973 fatal heart attacks per day.

Energy Comparisons.

37. Each hour of running requires 4×10^6 joules of energy, and the metabolism of each bar of candy releases 10^6 joules of energy, so a four hour run would require the energy of

$$4 \text{ hrs} \times \frac{4 \times 10^6 \text{ joules/hr}}{10^6 \text{ joules/candy bar}},$$

i.e., 16 candy bars.

39. Each kilogram of coal, when burned, releases 1.6×10^9 joules of energy, and the fission of one kilogram of uranium-235 releases 5.6×10^{13} joules of energy, so since

$$\frac{5.6 \times 10^{13} \text{ joules}}{1.6 \times 10^9 \text{ joules}} = 3.5 \times 10^4,$$

the fission of uranium-235 releases 35,000 times more energy than burning coal, kilogram by kilogram.

41. Since the fusion of hydrogen in 1 liter of water releases 7×10^{13} joules of energy, and an average home uses 5×10^7 joules of electrical energy per day, the amount of water required to generate (via fusion) the needs of such a home would be

$$\frac{5 \times 10^7 \text{ joules/day}}{7 \times 10^{13} \text{ joules/liter}},$$

which is about 0.7 of one millionth of a liter of water per day.

43. The fission of one kilogram of uranium-235 releases 5.6×10^{13} joules of energy, and the U.S. annual energy consumption is 1×10^{20} joules, so the energy needs of the U.S. would be met by

$$\frac{1 \times 10^{20} \text{ joules/year}}{5.6 \times 10^{13} \text{ joules/kilogram}},$$

or, about 1.8 million kilograms of uranium-235 per year.

45. Scale Ratio.

a. If 1 cm on the map represents 1 km = 10^3 m = 10^5 cm on the ground, the scale ratio for this map is 1 to 10,000.

b. If 2 inches on the map represents 0.5 mile on the ground, then as 1 mile is 5280 ft = 5280×12 in = 63,360 in, we see that 2 inches on the map represents 31,680 inches on the ground. Hence, 1 inch on the map represents 15,840 inches on the ground, and the scale ratio for this map is 1 to 15,840.

c. If 5 cm on the map represents 100 km = 100×10^5 cm = 10^7 cm on the ground, the scale ratio for this map is 5 to 10,000,000, which is 1 to 2,000,000.

d. If 1 ft on the map represents 100 meters = 100×3.28 ft = 3280 feet on the ground, the scale ratio for this map is 1 to 3280.

47. Scale Model Solar System.

We are given that the scale ratio for our model should be 1 to 10 billion ($= 10^{10}$). Thus, to find scaled sizes or distance, we divide the real sizes or distances by 10^{10}. Because the data table gives so many sizes and distances to put on this scale, it's easiest to look for general rules.

Note that many of the planet diameters are on the order of 10,000 km (104 km). On the 1 to 10 billion scale, a diameter of 10^4 km becomes a scaled diameter of $\frac{10^4 \text{ km}}{10^{10}} = 10^{-6}$ km, which is easier to visualize in millimeters. Since 1 km equals 10^6 mm, we see that 10^{-6} km equals one millimeter; thus the scale of 1 to 10 billion is equivalent to a scale where 1 mm corresponds to 10,000 km.

Now it is easy to convert the real diameters to scaled diameters. The earth's real diameter of about 12,760 km becomes a scaled diameter of about 1.3 mm, and Jupiter's real diameter of about 143,000 km becomes a scaled diameter of about 14.3 mm (which equals 1.43 cm). The second column in the table here shows the Model (scaled) diameters for the nine planets.

Object	Model Diameter	Model Distance
Mercury	0.5 mm	6 m
Venus	1.2 mm	11 m
Earth	1.3 mm	15 m
Mars	0.7 mm	23 m
Jupiter	14.3 mm	78 m
Saturn	12.0 mm	143 m
Uranus	5.2 mm	287 m
Neptune	4.8 mm	450 m
Pluto	0.2 mm	590 m

We can apply a similar argument to the distances. A real distance of 100 million km (10^8 km) becomes a scaled distance of $\frac{10^8 \text{ km}}{10^{10}} = 10^{-2}$ km, i.e., 10 meters (since there are $1000 = 10^3$ meters in one kilometer). Thus the 1 to 10 billion scale means that 10 m corresponds to 100 million km. Now, the Earth's real distance of about 150 million km from the Sun becomes a scaled distance of 15 meters, and Pluto's real distance of about 5,900 million km from the Sun becomes a scaled distance of 590 meters. This explains the third column above.

Putting things in context, it should be clear that "space" is pretty empty. For example, if Neptune were the size of the tip of a ballpoint pen, it would be over four football field lengths away from the sun!

49. The Amazing Amazon.

If the daily discharge of 4.5 trillion gallons can supply all U.S. households for five months, then $4.5/5 = 0.9$ trillion gallons can supply all U.S. households for one month. Assuming roughly 100 million U.S. households, this translates to $\frac{0.9\times10^{12}}{100\times10^6} = 9000$ gallons per household per month. This is reasonable, being $\frac{9000}{30} = 300$ gallons per household per day, and the actual per household usage for water is about 375 gallons per day.

51. Stellar Corpses: White Dwarfs and Neutron Stars.

a. We are told that the sun in its white dwarf incarnation will resemble a sphere of radius 6400 kilometers $= 6.4 \times 10^8$ centimeters, and have a mass of 2×10^{30} kilograms. Recall that a sphere of radius r has volume $\frac{4}{3}\pi r^3$. Since density is mass per unit volume, we compute:

$$\frac{2 \times 10^{30} \text{ kg}}{\frac{4}{3}\pi \times (6.4 \times 10^8 \text{ cm})^3} = 1.8214 \times 10^3 \frac{\text{kg}}{\text{cm}^3},$$

getting about 1800 kg/cm^3.

b. If a teaspoon measures about 4 cubic centimeters, then the mass of a teaspoon of the white dwarf would be roughly:

$$4\text{cm}^3 \times 1800\frac{\text{kg}}{\text{cm}^3} = 7200 \text{ kg},$$

which is roughly the mass of a large car.

c. We are dealing with a neutron star of radius 10 kilometers having a mass of $1.4 \times 2 \times 10^{30}$ kilograms, i.e., 2.8×10^{30} kilograms. Since 10 kilometers $= 10 \times 10^5 = 10^6$ centimeters, to find the density of the neutron star we compute:

$$\frac{2.8 \times 10^{30} \text{ kg}}{\frac{4}{3}\pi \times (10^6 \text{ cm})^3} = 6.68 \times 10^{11} \frac{\text{kg}}{\text{cm}^3}.$$

One cubic centimeter of this weighs about 668 billion kilograms, which is over ten times the total mass of Mount Everest (5×10^{10} kg, or, 50 billion kilograms).

Sampling Problems.

53. A stack of 10 pennies is roughly 1.5 cm high, so the thickness of one penny is approximately $1.5/10 = 0.15$ cm $= 1.5$ mm. A stack of 10 nickels is roughly 1.8 cm high, so the thickness of one nickel is approximately $1.8/10 = 0.18$ cm $= 1.8$ mm. A stack of 10 dimes is roughly 1.5 cm high, so the thickness of one dime is approximately $1.3/10 = 0.13$ cm $= 1.3$ mm. A stack of 10 quarters is roughly 1.5 cm high, so the thickness of one quarter is approximately $1.7/10 = 0.17$ cm $=$ 1.7 mm. (Similar calculations may be done using inches, but the metric answers are easier to work with and understand.) The thicknesses of these coins are of the same order of magnitude, so clearly it is to your advantage to take your height stacked in quarters, since the quarter is at least two and a half times as valuable as any of the other coins.

55. Solutions will vary.

Unit 3C Dealing with Uncertainty

Review of Rounding.

1. a. 2.

b. 15.

c. 779.

d. 4.

e. 14.

f. 235.

g. 2000.

h. 89.

i. -14.

3. Counting Significant Digits. Notice that the units attached to each number do not affect the number of significant digits.

a. The number 96.2 has three significant digits, because all digits are non-zero. It is precise to the nearest tenth (0.1) of a km/hr.

b. The number 100.020 has six significant digits, because the rightmost zero is usually unnecessary, so the fact that it is included makes it significant. It is precise to the nearest thousandth (0.001) of a second.

c. The number 0.00098 has two significant digits, because it can be written 9.8×10^4. It is precise to the nearest 0.00001 mm.

d. The number 0.0002020 has four significant digits, because it can be written 2.020×10^{-4} and the rightmost zero is usually unnecessary, so the fact that it is included makes it significant. It is precise to the nearest 0.0000001 meters.

e. The number 300,000 has only one significant digit, because it can be written 3×10^5. It is precise to the nearest 100,000.

f. The number 3×10^5 has only one significant digit. Since it equals 300,000, it is precise to the nearest 100,000.

Sources of Error.

5. There is likely to be random error due to miscounting because the intersection is busy, as well as systematic error (a particular person counting may well have their own bias) due to interpretation of what counts as a sports utility vehicle.

7. There is likely to be random error due to miscalculation of the average, and systematic error due to underreporting of incomes on tax returns.

9. There is likely to be random error due to errors in reading the scale, but no systematic error since the scale is well calibrated.

11. There is likely to be random error due

to miscounting.

13. Tax audit. The errors referred to in (1) are likely to be random error; those in (2) are more likely to be systematic (probably due to underreporting).

15. Safe Air Travel. An altimeter set to 2500 feet in a city that is actually at 5280 feet will read $5280 - 2500 = 2780$ feet too low at takeoff and throughout the subsequent flight. This is a systematic error.

Absolute and Relative Errors.

17. The absolute error is $\$19.00 - \$18.50 = \$0.50$; the relative error $\frac{\$0.50}{\$18.50} = 0.027 = 2.7\%$.

19. The absolute error is $\$48 - \$65 = -\$17$; the relative error $-\frac{\$17}{\$65} = -0.2615 = -26.15\%$.

21. The absolute error is 5.25 cups $- 5.5$ cups $= -0.25$ cups; the relative error is $-\frac{0.25 \text{ in}}{5.5 \text{ in}} = -0.04545 = -4.545\%$.

23. The absolute error is 11.7 mi $- 12$ mi $= -0.3$ mi; the relative error $-\frac{0.3 \text{ mi}}{12 \text{ mi}} = -0.025 = -2.5\%$.

Accuracy and Precision.

25. The tape measure is more accurate because it got closer to your true height than the laser did; but the laser is more precise, as it claimed to gauge your height to a greater degree of precision than the tape did.

27. The digital scale is more precise and more accurate than the clinical scale: it measures to a greater precision and got closer to your true weight than the tape measure did.

Believable Facts?

29. Random and systematic errors could be present. The claim is not believable with the given precision, even if it is an official census count.

31. Random and systematic errors could be present. The claim is believable with the given precision; it could be a reasonable projection.

33. Random and systematic errors could be present. The claim is believable with the given precision; presumably it's based on actual averaged measurements, rounded to the nearest degree.

35. Random and systematic errors could be present; the claim is not believable with the given precision, no matter what data it is based on.

37. Note that 140 liters $-$ 1.09 liters $=$ 138.91 liters, but since the volume of 140 liters is the least precise (to the nearest ten), we should round this difference to the nearest ten also, yielding 140 liters (again!).

39. The weight 9.7 kg has two significant digits, whereas 165 kg has three. The product of the given numbers is easily checked to be 1600.5 kg^2, but this should be rounded to two significant digits, namely 1600 kg^2.

41. Note that 36 miles $+$ 2.2 miles $=$ 38.2 miles, but since the first given distance was the least precise (to the nearest mile only), we should round the sum to the nearest mile also, yielding 38 miles.

43. The figure \$112.4 million has four significant digits, whereas 480,000 residents has only two. The appropriate quotient of these numbers is \$234.17 per resident, but rounded to two significant digits the answer should be listed as \$230 per resident.

Unit 3D How Numbers Deceive: Polygraphs, Mammograms, and More.

1. Batting Percentages.

In the first half of the season, Josh's batting average was $\frac{50}{150} = 0.333$, and Jude's batting average was $\frac{10}{50} = 0.200$, so Josh had the higher batting average. In the second half of the season, Josh's batting average was $\frac{35}{70} = 0.500$, and Jude's batting average was $\frac{70}{150} = 0.467$, so Josh had the higher batting average here too. However, combining the figures for both halves of the season, Josh's overall batting average was $\frac{50+35}{150+70} = 0.386$, and Jude's overall batting average was $\frac{10+70}{50+150} = 0.400$. Thus, Jude had the higher batting average overall. These results illustrate Simpson's paradox because Jude had the higher batting average overall despite the fact that Josh had the higher average in each half of the season considered separately.

3. Test Scores.

a. New Jersey had higher scores (283, 252 respectively) than Nebraska in both the white and non-white categories, yet Nebraska had the higher overall average (277) across both racial categories.

b. The percentage of non-whites in Nebraska is much less than in New Jersey.

c. If 87% of Nebraska students were white, with an average of 281, and the remaining 13% were non-white, with an average of 250, then assuming that N Nebraska students took the test, $0.87 \times N$ of them were white with scores (on average) of 281, whereas $0.13 \times N$ of them were non-white with scores (on average) of 250. Thus, their overall average score was

$$\frac{(0.87 \times N \times 281) + (0.13 \times N \times 250)}{N}$$

$$= (0.87 \times 281) + (0.13 \times 250) = 276.97.$$

d. If 66% of New Jersey students were white, with an average of 283, and the remaining 34% were non-white, with an average of 252, then assuming that M New Jersey students took the test, $0.66 \times M$ of them were white with scores (on average) of 283, whereas $0.34 \times M$ of them were non-white with scores (on average) of 252. Thus, their overall average score was

$$\frac{(0.66 \times M \times 283) + (0.34 \times N \times 252)}{M}$$

$$= (0.66 \times 283) + (0.34 \times 252) = 272.46.$$

e. Although Nebraska had lower scores in both categories, it came out ahead overall.

5. Tuberculosis Deaths.

a. In NYC, the death rate for whites was $\frac{8400}{4675000}$ or about 0.18%, and for non-whites it was $\frac{500}{92000}$ or about 0.54%. The overall death rate was $\frac{8400+500}{4675000+92000}$ or about 0.19%,

b. In Richmond, the death rate for whites was $\frac{130}{81000}$ or about 0.16%, and for non-whites it was $\frac{160}{47000}$ or about 0.34%. The overall death rate was $\frac{130+160}{81000+47000}$ or about 0.23%,

c. The death rate for both whites and non-whites in New York was higher than in Richmond; yet the overall death rate was higher in Richmond than in New York. This occurred here because the percentage of non-whites in New York was significantly less than that in Richmond.

7. Hiring Statistics.

Consider the men who applied: their acceptance rate for white-collar positions was 15%, their acceptance rate for blue-collar positions was 75%, and their overall acceptance rate was

$$\frac{(0.15)(200) + (0.75)(400)}{200 + 400},$$

or 55%.

Next, consider the women who applied: their acceptance rate for white-collar positions was 20%, and their acceptance rate for blue-collar positions was 85% – both higher than the corresponding rates for men. Yet the women's overall acceptance rate was

$$\frac{(0.20)(200) + (0.85)(100)}{200 + 100},$$

or about 41.67%,

Within each category—white-collar and blue-collar positions—the percentage of women hired was greater than the percentage of men hired, but overall, the hiring rate for men was higher than for women. This paradox can be resolved by noting that the number of blue-color jobs ($455 - 70 = 385$) was over five times greater than the number of white-collar jobs (70).

9. Basketball Records.

a. Spelman's home game record is $\frac{10}{10+19}$ or about 34.48% wins, and away game record is $\frac{12}{12+4}$ or 75% wins. This compares favorably with Morehouse, with a home game record of $\frac{9}{9+19}$ or about 32.14% wins, and an away game record of $\frac{56}{56+20}$ or about 73.68% wins. Go Spelman!!

b. Overall, Morehouse won $\frac{9+56}{9+19+56+20}$ or 62.5% of the time, whereas Spelman won $\frac{10+12}{10+19+12+4}$ or 48.89% of the time.

c. Since teams are generally rated on their overall record, Morehouse sadly has a better claim to superiority here.

11. Polygraph Test.

a. If the drug use rate is 1%, then $(0.01)(2000) = 20$ people actually use drugs, and $2000 - 20 = 1980$ do not. Next, 90% of users test positive, i.e., are deemed by the polygraph to have lied, so $(0.90)(20) = 18$ of the 20 users test positive, and $20 - 18 = 2$ do not. Similarly, 90% of the nonusers test negative, i.e., are deemed to have told the truth, this accounts for $(0.90)(1980) = 1782$ people. The remaining $1980 - 1782 = 198$ nonusers test positive. Overall, $18 + 198 = 216$ people are deemed to have lied, and $2 + 1782 = 1784$ people are deemed to have told the truth.

b. All 216 people who test positive are accused of lying, but only 18 of these were actually lying. The other 198 were telling the truth: $\frac{198}{216} = 91.67\%$ were falsely accused.

c. The polygraph claims that 1784 people were telling the truth, of whom 1782 really were; the other 2 were lying; thus, $\frac{1782}{1784}$ or about 99.89% of those found telling the truth were actually telling the truth.

13. HIV risks.

a. The HIV incidence rate in the general population is $\frac{57+3}{20000} = 0.003$, which is 0.3%. The HIV incidence rate in the "at risk" population is $\frac{475+25}{5000} = 0.100$, which is 10%. The proportion of true positives in the general population is $\frac{57}{57+3} = 0.95$. The proportion of true negatives in the general population is $\frac{18943}{997+18943} = 0.95$. The proportion of true positives in the "at risk" population is $\frac{475}{475+25} = 0.95$. The proportion of true negatives in the "at risk" population is $\frac{4275}{4275+225} = 0.95$. Thus the detection rates are 95% in all categories.

b. As just seen, 95% of those "at risk" are true positives, i.e., 95% of those "at risk" who have HIV test positive. However, only about 67.9% of those "at risk" who test positive have HIV, because $\frac{475}{475+225} = 0.679$. These numbers are different because in the first case the 475 "at risk" positive testees who are infected are being compared to the 500 "at risk" infected people, whereas in the second case the same people are being compared to all 700 "at risk" positive testees.

c. The chances of the patient having HIV are 67.9%, which is greater than the overall incidence rate of 10%: testing positive does increase the chance of being infected HIV.

d. As seen earlier, 95% of the general population are true positives, i.e., 95% of the general population with HIV test positive. However, only about 5.4% of the general population who test positive have HIV, because $\frac{57}{57+997} = 0.054$. These numbers are different because in the first case the 57 positive testees in the general population who are infected are being compared to the 60 infected people in the general population, whereas in the second case the same 57 people are being compared to all 1054 positive testees in the general population.

e. The chances of the patient having HIV are 5.4%, which is greater than the overall incidence rate of 0.3%, but nowhere nearly as great as the "95% accuracy" figure suggests.

Unit 4A The Power of Compounding

Algebra Review.

1. a. Given $x - 4 = 6$, we add 4 to both sides to get $x = 6 + 4 = 10$.

b. Given $y + 5 = 10$, we subtract 5 from both sides to get $y = 10 - 5 = 5$.

c. Given $2z = 12$, we divide both sides by 2 to get $z = 12/2 = 6$.

d. Given $3z = 15$, we divide both sides by 3 to get $x = 15/3 = 5$.

e. Given $2x - 5 = 13$, we first add 5 to both sides to get $2x = 13 + 5 = 18$, then divide by 2 to get $x = 18/2 = 9$.

f. Given $3n + 4 = 13$, we first subtract 4 from both sides to get $3n = 13 - 4 = 9$, then divide by 3 to get $n = 9/3 = 3$.

g. Given $4x + 8 = 24$, we first subtract 8 from both sides to get $4x = 24 - 8 = 16$, then divide by 4 to get $n = 16/4 = 4$.

h. Given $5 - 2w = 9$, we first subtract 5 from both sides to get $5 - 2w - 5 = 9 - 5 = 4$, i.e., $-2w = 4$, then divide by -2 to get $w = 4/(-2) = -2$.

Simple vs. Compound Interest.

3. Since Yancy is earning simple interest at an annual rate of 5%, her initial investment of \$500 grows by \$25 at the end of each year. After 5 years, her balance is up to \$625, which is a \$125 increase in absolute terms. In relative terms, her balance has increased by $\frac{\$125}{\$500} = 0.25$, i.e., 25%.

Samantha, on the other hand, invests the same amount of money at 5% compound interest. After the first year she too gets \$25 in interest and has a balance of \$525. However after the second year, she gets $0.05 \times \$525 = \$26.25$ in interest, resulting in a new balance of \$551.25. After the third year, she gets $0.05 \times \$551.25 = \$27.56$ in interest, resulting in a new balance of \$578.81. Similarly, after the fourth and fifth years she gets more "interest on her interest," as tabulated below, ending up with a balance of \$638.14.

	Yancy		*Samantha*	
Year	*Interest*	*Balance*	*Interest*	*Balance*
0	-	\$500	-	\$500
1	\$25	\$525	\$25	\$525
2	\$25	\$550	\$26.25	\$551.25
3	\$25	\$575	\$27.56	\$578.81
4	\$25	\$600	\$28.94	\$607.75
5	\$25	\$625	\$30.39	\$638.14

After 5 years, Samantha's balance is up by \$138.14, which in relative terms is $\frac{\$138.14}{\$500} = 0.276$, i.e., a 27.6% increase.

Compound Interest.

5. If we invest \$2000 at an APR of 3% for 10 years, we accumulate a balance of $\$2000 \times (1 + 0.03)^{10} = \$2687.83$.

7. If we invest \$30,000 at an APR of 7% for 25 years, we accumulate a balance of $\$30,000 \times (1 + 0.07)^{25} = \$162,822.98$.

9. If we invest \$10,000 at an APR of 6% for 25 years, we accumulate a balance of $\$10,000 \times (1 + 0.06)^{25} = \$42,918.71$.

Small Rate Differences.

11. After 10 years Chang has $\$500 \times (1 + 0.035)^{10} = \$705.30$; after 30 years he has $\$500 \times (1 + 0.035)^{30} = \$1403.40$. After 10 years Kio has $\$500 \times (1+0.0375)^{10} = \$722.52$; after 30 years she has $\$500 \times (1+0.0375)^{30} =$

\$1508.74. Hence, Kio has \$17.22 more than Chang after 10 years, and since $\frac{\$17.22}{\$705.30} = 0.024$, Kio's balance is 2.4% higher than Chang's.

After 30 years the change is more dramatic: Kio has \$105.34 more than Chang, and since $\frac{\$105.34}{\$1403.40} = 0.075$, Kio's balance is 7.5% higher than Chang's. Consequently, a small interest rate difference gets magnified over time.

13. Investing \$1000 at an APR of 5.5% compounded monthly for 10 years yields: $A = \$1000(1 + \frac{0.055}{12})^{120} = \$1731.08$.

15. Investing \$5000 at an APR of 7.3% compounded quarterly for 20 years yields: $A = \$5000(1 + \frac{0.073}{4})^{80} = \$21,248.27$.

17. Investing \$1000 at an APR of 7% compounded monthly for 10 years yields: $A = \$1000(1 + \frac{0.07}{12})^{180} = \$2848.95$.

19. Investing \$5000 at an APR of 6.2% compounded quarterly for 30 years yields: $A = \$5000(1 + \frac{0.062}{4})^{120} = \$31,663.68$.

Annual Percentage Yield (APY).

21. Investing \$1000 for 1 year at an APR of 6.5% compounded daily yields $A = \$1000(1 + \frac{0.065}{365})^{365} = \$1067.15$, which results in an APY of $\frac{\$67.15}{\$1000} = 0.06715$, i.e., 6.72%.

23. Investing \$1000 for 1 year at an APR of 3.25% compounded monthly yields $A = \$1000(1 + \frac{0.0325}{12})^{12} = \$1032.99$, which results in an APY of $\frac{\$32.99}{\$1000} = 0.03299$, i.e., 3.30%.

25. Comparing Annual Yields. If we invest \$1000 for 1 year at an APR of 6.6% compounded quarterly, we end up with $A = \$1000(1 + \frac{0.066}{4})^{4} = \$1067.65$, which results in an APY of $\frac{\$67.65}{\$1000} = 0.06765$, i.e., 6.77%. If the interest is compounded monthly, we get $A = \$1000(1 + \frac{0.066}{12})^{12} = \$1068.03$, and an APY of 6.80%. Daily compounding leads to $A = \$1000(1 + \frac{0.066}{365})^{365} = \$1068.22$, and an APY of 6.82%.

Compounding monthly as opposed to quarterly increases the APY noticably, but increasing the frequency of compounding from monthly to daily has a smaller effect.

27. Rates of Compounding. For Account 1, after Y years the balance is given by $\$1000(1 + 0.055)^{Y}$, and the interest is that amount less the initial \$1000. Letting $Y = 1, 2, \ldots, 10$ yields half of the table below (in which we rounded to the nearest dollar).

	Account 1		*Account 2*	
Year	*Interest*	*Balance*	*Interest*	*Balance*
0	-	\$1000	-	\$1000
1	\$55	\$1055	\$57	\$1057
2	\$58	\$1113	\$59	\$1116
3	\$61	\$1174	\$63	\$1179
4	\$65	\$1239	\$67	\$1246
5	\$68	\$1307	\$70	\$1317
6	\$72	\$1379	\$74	\$1391
7	\$76	\$1455	\$79	\$1470
8	\$80	\$1535	\$83	\$1553
9	\$84	\$1619	\$87	\$1640
10	\$89	\$1708	\$93	\$1733

For Account 2, after Y years the balance is $\$1000(1 + \frac{0.055}{365})^{365Y}$, and the interest is that amount less the initial \$1000. By varying Y, we obtain the other half of the table.

After 10 years, Account 1 has increased in value by \$708, which in relative terms is 70.8%. Account 2 has increased in value by \$733, which in relative terms is 73.3%.

Continuous Compounding. In these problems, we use the formula $A = Pe^{APR\times Y}$.

29. If we invest \$1000 for 1 year at 4% compounded continuously, we end up with $A = \$1000e^{0.04\times 1} = \$1040.81$, and an APY of $\frac{\$40.81}{\$1000} = 0.04081$, i.e., 4.08%. Investing for 5 years we get $A = \$1000e^{0.04\times 5} = \$1221.40$, and investing for 20 years we get $A = \$1000e^{0.04\times 20} = \$2225.54$.

31. If we invest \$10000 for 1 year at 6% compounded continuously, we end up with $A = \$10000e^{0.06\times 1} = \$10,618.37$, and an APY of $\frac{\$618.37}{\$10000} = 0.06184$, i.e., 6.18%. Investing for 5 years we get $A = \$10000e^{0.06\times 5} = \$13,498.59$, and investing for 20 years we get $A = \$10000e^{0.06\times 20} = \$33,201.17$.

33. If we invest \$5000 for 1 year at 6.5% compounded continuously, we end up with $A = \$5000e^{0.065\times 1} = \$5335.80$, and an APY of $\frac{\$335.80}{\$5000} = 0.06716$, i.e., 6.72%. Investing for 5 years we get $A = \$5000e^{0.065\times 5} = \$6920.15$, and investing for 20 years we get $A = \$5000e^{0.065\times 20} = \$18,346.48$.

35. Comparing Investment Plans. Bernard invests \$1600 at 4% compounded annually, so after Y years he has $A = \$1600(1+0.04)^Y$. For $Y = 5$ we get \$1946.64, and for $Y = 20$ we get \$3505.80.

Carla invests \$1400 at an APR of 5% compounded daily; after Y years she has $A = \$1400(1 + \frac{0.045}{365})^{365Y}$. For $Y = 5$ we get \$1797.60, and for $Y = 20$ we get \$3805.33.

Hence, Bernard has the higher accumulated balance after 5 years, but Carla comes out ahead after 20 years: Carla's higher APR and more frequent compounding allows her to more than make up for her smaller investment given sufficient time.

Planning Ahead with Compounding. In these problems, we use the formulae

$$P = \frac{A}{(1 + \frac{APR}{n})^{nY}} \quad \text{and} \quad P = \frac{A}{e^{APR\times Y}},$$

for interest compounded n times a year and continuously, respectively.

37. To end up with \$10,000 after 10 years at an APR of 9% compounded annually, you must start with $P = \frac{\$10000}{(1+0.09)^{10}} = \$4224.11$.

39. To end up with \$10,000 after 10 years at an APR of 9% compounded monthly, you must start with $P = \frac{\$10000}{(1+\frac{0.09}{12})^{120}} = \$4079.37$.

41. College Fund.

a. To get \$100,000 after 18 years at an APR of 9% compounded daily, you must start with $P = \frac{\$100000}{(1+\frac{0.06}{365})^{6570}} = \$33,962.57$.

b. To get \$100,000 after 18 years at an APR of 7.5% compounded continuously, you must start with $P = \frac{\$100000}{e^{1.35}} = \$25,924.03$.

c. To get \$100,000 after 18 years at an APR of 11% compounded monthly, you must start with $P = \frac{\$100000}{(1+\frac{0.11}{12})^{216}} = \$13,932.02$.

43. Your Bank Account.

Solutions will vary.

45. Finding Time Periods.

a. Investing \$1000 at an APR of 8% for Y years yields $A = 1000(1.08)^Y$. For $Y = 10$ we get $A = \$2158.92$, which falls quite a bit short short of tripling our \$1000. Trying $Y = 15$ we get $A = \$3172.17$, which more than triples our money. Trying $Y = 14$ we get $A = \$2937.19$, which is a bit short of our goal. However, since the interest is being compounded only once a year, we must settle for 15 years and an overshoot of our goal.

b. Investing \$1000 at an APR of 7% for Y years yields $A = 1000(1.07)^Y$. For $Y = 5$ we get $A = \$1402.55$, which is a bit short short of our desired \$1500. Trying $Y = 6$ we get $A = \$1500.73$, which is almost exactly what we wanted. Since the interest is being compounded only once a year, we settle for 6 years, and a tiny overshoot of our goal.

c. As above we seek the smallest whole number Y so that $A = 1000(1.07)^Y$ comes out to be \$100,000 or larger.

For $Y = 25$ we get $A = \$5427.43$, which is far short of our desired goal. $Y = 50$ yields $A = \$29,457.03$, which is still short. $Y = 75$ yields $A = \$159,876.01$, which is a considerable overshoot! Trying $Y = 65, 70, 68,$ and 69 in succession we find that 68 years is not enough, and 69 gives us \$106,532.14. Since the interest is being compounded only once a year, we must settle for 69 years.

Unit 4B Savings Plans

Review of Powers and Roots.

1. a. $4^3 = 4 \times 4 \times 4 = 64$.
b. $2^3 \times 2^5 = 2^{3+5} = 2^8 = 256$.
c. $3^6 \div 3^2 = 3^{6-2} = 3^4 = 81$.
d. $4^{-2} = \frac{1}{4^2} = \frac{1}{4\times 4} = \frac{1}{16}$.
e. $4^5 \times 4^{-2} = 4^{5-2} = 4^3 = 64$.
f. $5^3 \div 5^{-4} = 5^{3-(-4)} = 5^7 = 78,125$.

3. a. Given $x^2 = 100$, we get $x = \pm 10$.
b. Given $x^3 = 27$, we get $x = 3$.
c. Given $x^{\frac{1}{3}} = 2$, we get $x = 2^3 = 8$.
d. Given $x^{\frac{1}{5}} = 2$, we get $x = 2^5 = 32$.

Investment Plans. Recall the formula:

$$A = PMT \times \frac{(1 + \frac{APR}{n})^{nY} - 1}{\frac{APR}{n}}.$$

5. Depositing \$50 monthly for 40 years at 8% yields

$$A = \$50 \times \frac{(1 + \frac{0.08}{12})^{12\times 40} - 1}{\frac{0.08}{12}} = \$174,550.39.$$

This is more than 7 times the total deposits $\$50 \times 12 \times 40 = \$24,000$ made.

7. Depositing \$200 monthly for 18 years at 7% yields

$$A = \$200 \times \frac{(1 + \frac{0.07}{12})^{12\times 18} - 1}{\frac{0.07}{12}} = \$86,144.21.$$

This is about twice the total amount of deposits made over the 18 years, which is $\$200 \times 12 \times 18 = \$43,200$.

Who Comes Out Ahead?

9. Yolanda deposits \$100 monthly for 10 years at 5%, yielding

$$A = \$100 \times \frac{(1 + \frac{0.05}{12})^{12\times 10} - 1}{\frac{0.05}{12}} = \$15,528.23.$$

Overall, she deposits $\$100 \times 12 \times 10 = \$12,000$.

Zach deposits \$1200 once a year for 10 years at 5%, yielding

$$A = \$1200 \times \frac{(1 + \frac{0.05}{1})^{1\times 10} - 1}{\frac{0.05}{1}} = \$15,093.47.$$

Overall, he deposits $\$1200 \times 10 = \$12,000$.

Thus we see that although both people deposit the same amount of money overall, and have the same APR, Yolanda comes out ahead. This is because her interest is compounded more frequently.

11. Juan deposits \$200 monthly for 10 years at 6%, yielding

$$A = \$200 \times \frac{(1 + \frac{0.06}{12})^{12\times10} - 1}{\frac{0.06}{12}} = \$32,775.87.$$

Overall, he deposits $\$200 \times 12 \times 10 = \$24,000$.

Maria deposits \$2500 once a year for 10 years at 6.5%, yielding

$$A = \$2500 \times \frac{(1 + \frac{0.065}{1})^{1\times10} - 1}{\frac{0.065}{1}} = \$33,736.06.$$

Overall, she deposits $\$2500 \times 10 = \$25,000$. Maria comes out just ahead of Juan, despite her less frequent compounding. This is because her APR is higher than his, and overall she deposits a bit more than he does.

Investment Planning. We will need the inverted savings plan formula, which gives the PMT in terms of the amount A saved:

$$PMT = \frac{A \times \frac{APR}{n}}{(1 + \frac{APR}{n})^{nY} - 1}.$$

13. To create a college fund worth \$150,000 by making monthly deposits for 18 years, assuming an APR of 7.5%, you must deposit

$$\frac{\$150,000 \times \frac{0.075}{12}}{(1 + \frac{0.075}{12})^{12\times18} - 1} = \$329.96$$

each month.

15. To save up \$10,000 to buy a car, by making monthly deposits for 3 years at an APR of 5.5%, each month you must deposit

$$\frac{\$10,000 \times \frac{0.055}{12}}{(1 + \frac{0.055}{12})^{12\times3} - 1} = \$256.13.$$

Comparing Investment Plans.

17. If you deposit \$50 monthly for 15 years at 7%, you will end up with

$$A = \$50 \times \frac{(1 + \frac{0.07}{12})^{12\times15} - 1}{\frac{0.07}{12}} = \$15,848.11,$$

which is considerably short of your desired goal of \$50,000. This investment plan is far from adequate.

19. If you deposit \$100 monthly for 15 years at 6%, you will end up with

$$A = \$100 \times \frac{(1 + \frac{0.06}{12})^{12\times15} - 1}{\frac{0.06}{12}} = \$29,081.87,$$

which is still short of your desired goal of \$50,000. This investment plan is far from adequate.

21. Comfortable Retirement. First, we need to figure out how much you must save by the time you reach 60. In order to draw an annual income of \$50,000 from then on, without end, if we assume an APR of 8%, then the amount saved must satisfy $A \times 0.08 = \$50,000$. Hence, $A = \frac{\$50,000}{0.08} = \$625,000$. Next, in order to save \$625,000 after 30 years of monthly deposits at 8%, you must deposit

$$\frac{\$625,000 \times \frac{0.08}{12}}{(1 + \frac{0.08}{12})^{12\times30} - 1} = \$419.36$$

each month.

23. Since 100 shares of the stock in question cost you \$5500, and you sold the stock after five years for \$10,300, the total return was $\frac{10300-5500}{5500} = 0.8727 = 87.27\%$, and the annual return was

$$\left(\frac{10300}{5500}\right)^{\frac{1}{5}} - 1 = 0.1337 = 13.37\%.$$

25. Since the shares cost you $5500, and you receive $11,300 for them after 10 years, the total return was $\frac{11300-5500}{5500} = 1.0545 =$ 105.45%, and the annual return was

$$\left(\frac{11300}{5500}\right)^{\frac{1}{10}} - 1 = 0.0747 = 7.47\%.$$

27. Since the shares cost you $5000, and you only receive $3000 for them after 5 years, the total return was $\frac{3000-5000}{5000} = -0.4000 =$ -40.00% (a loss), and the annual return was

$$\left(\frac{3000}{5000}\right)^{\frac{1}{5}} - 1 = -0.0971 = -9.71\%.$$

29. Since the shares cost you $7500, and you sold the stock after 8 years for $12,600, the total return was $\frac{12600-7500}{7500} = 0.6800 =$ 68.00%, and the annual return was

$$\left(\frac{12600}{7500}\right)^{\frac{1}{8}} - 1 = 0.0670 = 6.70\%.$$

31. Solutions will vary.

Unit 4C Loan Payments, Credit Cards, and Mortgages

1. Loan Terminology.

a. For this loan, the starting principal is $40,000, the APR is 7%, the number of payments per year is 12, the loan term is 20 years, and the payment amount is $310.

b. Overall, you will make $12 \times 20 = 240$ payments, and the total payments will be $\$310 \times 240 = \$74,400$.

c. $40,000 of the payments pays off the principal, and the remaining $\$74,400 - \$40,000 = \$34,400$ is all interest.

Loan Payments. Recall the formula:

$$PMT = \frac{P \times \frac{APR}{n}}{1 - \left(1 + \frac{APR}{n}\right)^{-nY}}.$$

3. a. A 20-year loan of $25,000 at an APR of 10% requires monthly payments of

$$\frac{\$25,000 \times \frac{0.10}{12}}{1 - \left(1 + \frac{0.10}{12}\right)^{(-12\times20)}} = \$241.26.$$

b. The total payments over the term of the loan will be $\$241.26 \times 12 \times 20 = \$57,902.40$.

c. $25,000 of the payments pays off the principal, and the remaining $\$57,902.40 - \$25,000 = \$32,902.40$ is all interest.

5. a. A 30 year mortgage of $150,000 at an APR of 7.5% requires monthly payments of

$$\frac{\$150,000 \times \frac{0.075}{12}}{1 - \left(1 + \frac{0.075}{12}\right)^{(-12\times30)}} = \$1048.82.$$

b. The total payments over the term of the mortgage will be $\$1048.82 \times 12 \times 30 =$ $\$377,575.20$.

c. $150,000 of the payments pays off the principal, and the remaining $\$377,575.20 -$ $\$150,000 = \$227,575.20$ is all interest.

7. a. A 30 year mortgage of $100,000 at an APR of 8.5% requires monthly payments of

$$\frac{\$100,000 \times \frac{0.085}{12}}{1 - \left(1 + (\frac{0.085}{12}\right)^{(-12\times30)}} = \$768.91.$$

b. The total payments over the term of the mortgage will be $\$768.91 \times 12 \times 30 =$ $\$276,807.60$.

c. $25,000 of the payments pays off the principal, and the remaining $\$276,807.60 -$ $\$100,000 = \$176,807.60$ is all interest.

9. a. A 3-year loan of \$5000 at an APR of 12% requires monthly payments of

$$\frac{\$5000 \times \frac{0.12}{12}}{1-\left(1+\frac{0.12}{12}\right)^{(-12\times 3)}} = \$166.07.$$

b. The total payments over the term of the loan will be $\$166.07 \times 12 \times 3 = \$5978.52$.

c. \$5000 of the payments pays off the principal, and the remaining $\$5978.52 - \$5000 = \$978.52$ is all interest.

11. a. A 15-year loan of \$50,000 at an APR of 8% requires monthly payments of

$$\frac{\$50,000 \times \frac{0.08}{12}}{1-\left(1+\frac{0.08}{12}\right)^{(-12\times 15)}} = \$477.83.$$

b. The total payments over the term of the loan will be $\$477.83 \times 12 \times 15 = \$86,009.40$.

c. \$50,000 of the payments pays off the principal, and the remaining $\$86,009.40 - \$50,000 = \$36,009.40$ is all interest.

Principal and Interest Payments.

13. A 30 year mortgage of \$100,000 at an APR of 8.5% requires monthly payments of

$$\frac{\$100,000 \times \frac{0.085}{12}}{1-\left(1+\frac{0.085}{12}\right)^{(-12\times 30)}} = \$768.91.$$

Note that the monthly interest rate here is $\frac{0.085}{12} = 0.0070833$. For a \$100,000 starting loan principal, the interest due at the end of the first month is $0.0070833 \times \$100,000 = \$708.33$. So of the \$768.91 first monthly payment, $\$768.91 - \$708.33 = \$60.58$ goes towards the principal. This effectively reduces the loan to $\$100,000.00 - \$60.58 = \$99,939.42$.

The interest due at the end of the second month is $0.0070833 \times \$99,939.42 = \$707.90$. So of the \$768.91 second monthly payment, $\$768.91 - \$707.90 = \$61.01$ goes towards the principal. This effectively reduces the loan to $\$99,939.42 - \$61.01 = \$99,878.41$.

End of Month	*Interest paid*	*Toward principal*	*New principal*
1	\$708.33	\$60.58	\$99,939.42
2	\$707.90	\$61.01	\$99,878.41
3	\$707.47	\$61.44	\$99,816.97

A similar argument shows that an additional \$61.44 is paid towards the principal at the end of the third month, and the loan is thereby reduced to \$99,816.97.

15. Choosing an Auto Loan. A 3-year loan of \$10,000 at an APR of 7% requires monthly payments of

$$\frac{\$10,000 \times \frac{0.07}{12}}{1-\left(1+\frac{0.07}{12}\right)^{(-12\times 3)}} = \$308.77,$$

which is much more than you can afford. Borrowing the same amount of money for 4 years at 7.5% requires monthly payments of

$$\frac{\$10,000 \times \frac{0.075}{12}}{1-\left(1+\frac{0.075}{12}\right)^{(-12\times 4)}} = \$241.79,$$

which is still beyond the reach of your budget. A 5-year loan of \$10,000 at 8% requires monthly payments of

$$\frac{\$10,000 \times \frac{0.08}{12}}{1-\left(1+\frac{0.08}{12}\right)^{(-12\times 5)}} = \$202.76,$$

which is affordable. This is the only loan which meets your needs.

Credit Card Debt.

17. If you pay off a credit card debt of \$2500 in 1 year at an APR of 18%, your monthly payments will be

$$\frac{\$2500 \times \frac{0.18}{12}}{1-\left(1+\frac{0.18}{12}\right)^{(-12\times 1)}} = \$229.20,$$

and the total payments will be $12 \times$ \$229.20 = \$2750.40.

19. If you pay off a credit card debt of \$2500 in 3 years at an APR of 21%, your monthly payments will be

$$\frac{\$2500 \times \frac{0.21}{12}}{1-\left(1+\frac{0.21}{12}\right)^{(-12\times 3)}} = \$94.19,$$

and the total payments will be $36 \times$ \$94.19 = \$3390.84.

21. Credit Card Debt. Suppose we make monthly payments of \$200 towards a balance of \$1200 at an APR of 18%, which is equivalent to a monthly interest rate of 1.5%. At the end of each month, you reduce your balance by \$200 but also increase it by \$75, plus the interest on the previous month's balance.

M	*Payment*	*Expenses*	*Interest*	*Balance*
0	-	-	-	\$1200.00
1	\$200	\$75	\$18.00	\$1093.00
2	\$200	\$75	\$16.40	\$984.40
3	\$200	\$75	\$14.77	\$874.17
4	\$200	\$75	\$13.11	\$762.28
5	\$200	\$75	\$11.43	\$648.71
6	\$200	\$75	\$9.73	\$533.44
7	\$200	\$75	\$8.00	\$416.44
8	\$200	\$75	\$6.25	\$297.69
9	\$200	\$75	\$4.47	\$177.16
10	\$200	\$75	\$2.66	\$54.82

At the end of the first month your balance becomes: \$1200.00 – \$200 + \$75 + (1.5% × \$1200.00) = \$1093.00. At the end of the second month it's: \$1093.00 – \$200 + \$75 + (1.5% × \$1093.00) = \$984.49. At the end of the third month the balance is: \$984.40 – \$200 + \$75 + (1.5% × \$984.40) = \$874.17. Continuing in this way, we get the entries in the last column of the above table.

At the end of the tenth month the balance becomes: \$177.16 – \$200 + \$75 + (1.5% × \$177.16) = \$54.82. The eleventh payment of \$200 is more than we need to pay this off; a partial payment suffices.

23. Credit Card Woes. Suppose we make irregular payments ranging from zero to \$500 per month towards paying off a credit card debt of \$300, all the while incuring additional monthly expenses, as documented in the second and third columns of the table below. The APR is 18%, so that the monthly interest rate is 1.5%.

M	*Payment*	*Expenses*	*Interest*	*Balance*
0	-	-	-	\$300
1	\$300	\$175	\$4.50	\$179.50
2	\$150	\$150	\$2.69	\$182.19
3	\$400	\$350	\$2.73	\$134.92
4	\$500	\$450	\$2.02	\$86.94
5	0	\$100	\$1.30	\$188.24
6	\$100	\$100	\$2.82	\$191.06
7	\$200	\$150	\$2.87	\$143.93
8	\$100	\$80	\$2.16	\$126.09

At the end of the first month your balance becomes: \$300.00 – \$300 + \$175 + (1.5% × \$300.00) = \$179.50. At the end of the second month it's: \$179.50 – \$150 + \$150 + (1.5% × \$179.50) = \$182.19, and so on. At the end of the eighth month the balance becomes: \$143.93 – \$100 + \$80 + (1.5% × \$143.93) = \$126.09.

In spite of the fact that for 7 of the 8 months, expenses did not exceed payments, the initial balance has only been reduced by a bit more than half: the accumulating interest and continued spending counteracts much of our well-intentioned payments.

Fixed Rate Options.

25. With the first option, the monthly payments will be

$$\frac{\$100,000 \times \frac{0.08}{12}}{1 - \left(1 + \frac{0.08}{12}\right)^{(-12\times 30)}} = \$733.76,$$

and the total payments will be $12 \times 30 \times \$733.76 = \$264,153.60$.

With the second option, the monthly payments will be

$$\frac{\$100,000 \times \frac{0.075}{12}}{1 - \left(1 + \frac{0.075}{12}\right)^{(-12\times 15)}} = \$927.01,$$

and the total payments will be $12 \times 15 \times \$927.01 = \$166,861.80$.

Payments for the 15-year loan will be larger than for the 30-year loan; so you do need more money each month to make the payments. However, with the 15-year mortgage, you will pay the loan off sooner and also pay less overall: in the end, the 30-year mortgage costs over 50% more than the 15-year mortgage. If the higher payment is affordable, the first option is probably a better one. (Not considered here are the tax consequences of mortgage interest payments.)

27. With the first option, the monthly payments will be

$$\frac{\$120,000 \times \frac{0.0715}{12}}{1 - \left(1 + \frac{0.0715}{12}\right)^{(-12\times 30)}} = \$810.49,$$

and the total payments will be $12 \times 30 \times \$810.49 = \$291,776.40$.

With the second option, the monthly payments will be

$$\frac{\$120,000 \times \frac{0.0675}{12}}{1 - \left(1 + \frac{0.0675}{12}\right)^{(-12\times 15)}} = \$1061.89,$$

and the total payments will be $12 \times 15 \times \$1061.89 = \$191,140.20$.

Payments for the 15-year loan will be larger than for the 30-year loan; so you do need more money each month to make the payments. However, with the 15-year mortgage, you will pay the loan off sooner and also pay less overall: in the end, the 30-year mortgage costs over 50% more than the 15-year mortgage. If the higher payment is affordable, the first option is a better one.

Closing Costs.

29. With the first option, the monthly payments will be

$$\frac{\$80,000 \times \frac{0.08}{12}}{1 - \left(1 + \frac{0.08}{12}\right)^{(-12\times 30)}} = \$587.01,$$

and the closing costs of $1200.

With the second option, the monthly payments will be

$$\frac{\$80,000 \times \frac{0.075}{12}}{1 - \left(1 + \frac{0.075}{12}\right)^{(-12\times 30)}} = \$559.37,$$

the closing costs are = $1200, and we must also pay 2% of the principal (the points). Thus the total extra costs for the second option come to $\$1200 + (0.02 \times \$80,000) = \$1200 + \$1600 = \$2800$.

The total payout for the first option—which includes 30 years of 12 monthly payments as above—is $\$211,323.60 + \$1200 =$

$212,523.60. The total payout for the second option is $201,373.20 + $2800 = $204,173.20. The second loan, with its lower interest rate, is the better option, despite the points.

31. With the first option, the monthly payments will be

$$\frac{\$80,000 \times \frac{0.0725}{12}}{1-\left(1+\frac{0.0725}{12}\right)^{(-12\times 30)}} = \$545.74,$$

and closing costs of $1200, and we must also pay 1% of the principal (the points). Thus the total extra costs for the first option come to $1200+(0.01×$80,000) = $1200+$800 = $2000.

With the second option, the monthly payments will be

$$\frac{\$80,000 \times \frac{0.0675}{12}}{1-\left(1+\frac{0.0675}{12}\right)^{(-12\times 30)}} = \$518.88,$$

the closing costs are = $1200, and we must also pay 3% of the principal (the points). Thus the total extra costs for the second option come to $1200 + (0.03 × $80,000) = $1200 + $2400 = $3600.

The total payout for the first option—which includes 30 years of 12 monthly payments as above—is $196,466.40 + $2000 = $198,466.40. The total payout for the second option is $186,796.80 + $3600 = $190,396.80. The second loan, with its lower interest rate, is the better option, despite the additional points.

33. Accelerated Loan Payment.

a. A 20-year loan of $25,000 at an APR of 9% requires monthly payments of

$$\frac{\$25,000 \times \frac{0.09}{12}}{1-\left(1+\frac{0.09}{12}\right)^{(-12\times 20)}} = \$224.93.$$

b. If we pay off the loan in 10 years instead of 20 years, the payments increase to:

$$\frac{\$25,000 \times \frac{0.09}{12}}{1-\left(1+\frac{0.09}{12}\right)^{(-12\times 10)}} = \$316.69.$$

c. For the 20-year loan, the total payments are 12 × 20 × $224.93 = $53,983.20, and for the 10-year loan they are 12×10×$316.69 = $38,002.80.

35. ARM Rate Approximations. For the 7% fixed rate $150,000 loan, the interest payments in the first year will be approximately 7% × $150,000 = $10,500. For the 5% ARM loan, the interest payments in the first year will be approximately 5%×$150,000 = $7500. Thus the ARM loan will save you about $3000 in the first year, which comes out to $250 per month. If the ARM later rises to 8.5%, the annual interest will increase to roughly 8.5% × $150,000 = $12,750. This is $2250 *more* than the interest on the fixed rate loan, which is about $188 more per month.

37. How Much House Can You Afford? The loan payment formula

$$PMT = \frac{P \times \frac{APR}{n}}{1-\left(1+\frac{APR}{n}\right)^{-nY}}$$

can be turned around to yield:

$$P = PMT \times \frac{1-\left(1+\frac{APR}{n}\right)^{-nY}}{\frac{APR}{n}}.$$

In our case, since we can make monthly payments of \$500 on a 30-year loan at 9%, we can afford a loan principal of:

$$P = \$500 \times \frac{1 - \left(1 + \frac{0.09}{12}\right)^{-12\times 30}}{\frac{0.09}{12}},$$

or about \$62,000.

Now suppose we have the cash on hand to make a 20% down payment, this means that we must borrow the remaining 80%. Since we can afford to borrow \$62,000, we can afford to buy a house for which \$62,000 represents 80% of the cost. As can easily be checked, such a house costs \$77,500.

39. Student Loan Consolidation. a. The 15-year loan of \$10,000 at an APR of 8% requires monthly payments of

$$\frac{\$10,000 \times \frac{0.08}{12}}{1 - \left(1 + \frac{0.08}{12}\right)^{(-12\times 15)}} = \$95.57,$$

the 20-year loan of \$15,000 at an APR of 8% requires monthly payments of

$$\frac{\$15,000 \times \frac{0.085}{12}}{1 - \left(1 + \frac{0.085}{12}\right)^{(-12\times 20)}} = \$130.17,$$

and the 10-year loan of \$12,500 at an APR of 8% requires monthly payments of

$$\frac{\$12,500 \times \frac{0.09}{12}}{1 - \left(1 + \frac{0.09}{12}\right)^{(-12\times 15)}} = \$158.34.$$

b. The total payments for the 15-year loan are $\$95.57 \times 12 \times 15 = \$17,202.60$, the total payments for the 20-year loan are $\$130.17 \times 12 \times 20 = \$31,240.80$, and the total payments for the 10-year loan are $\$158.34 \times 12 \times 10 = \$19,000,80$. Adding these up we get \$67,444.20.

c. A 20-year loan of $\$10,000 + \$15,000 + \$12,500 = \$37,500$ at an APR of 8.5% requires monthly payments of

$$\frac{\$37,500 \times \frac{0.085}{12}}{1 - \left(1 + \frac{0.085}{12}\right)^{(-12\times 20)}} = \$325.43,$$

which amounts to $\$325.43 \times 12 \times 20 = \$78,103.20$ over the term of the loan.

There is something to be said for this consolidated loan: you make one simple payment each month to one bank, and overall you pay not too much more than if you stick with the three separate loans. On the other hand, while the three loans require you to pay \$384.08 (the sum of the three individual payments) monthly at first, this drops to \$225.74 (the sum of the payments for the 15- and 20-year loans) after 10 years, and again 5 years later, as two of the loans are paid off.

41. Solutions will vary.

Unit 5A Fundamentals of Statistics

Population and Sample.

1. Population: registered voters in California; sample: 1026 people selected for interviews; population parameters: approval ratings of candidates for all voters; sample statistics: approval ratings of candidates for those in sample.

3. Population: all new model computers; sample: one computer selected for measurements; population parameter: speed of all new model computers on specific tasks; sample statistic: speed of the selected computer on specific tasks.

5. Population: all adult Americans; sample: 1010 selected adults; population parameters: rankings of professions by all adult Americans; sample statistics: rankings of professions by those in sample.

Steps in a Study.

7. Step 1: population is all students at the school; goal is to determine annual pizza consumption. Step 2: Choose a representative sample. Step 3: Determine annual pizza consumption for those in sample. Step 4: Infer annual pizza consumption for all students. Step 5: Assess results; formulate conclusion.

9. Step 1: population is all adult American women; goal is to determine mean height of these women. Step 2: Choose a representative sample. Step 3: Determine mean height for those in sample. Step 4: Infer mean height for all adult American women. Step 5: Assess results; formulate conclusion.

11. Step 1: population is all alkaline batteries; goal is to determine mean lifetime of these batteries. Step 2: Choose a representative sample. Step 3: Determine mean lifetime of batteries in sample. Step 4: Infer mean lifetime of all alkaline batteries. Step 5: Assess results; formulate conclusion.

13. Representative Sample? The entire track team would of course be a representative—but unnecessarily large—sample! The seniors would not be a representative sample, as age could be an important factor. The putters and throwers, or sprinters, would not be representative either, as their high energy activities may result in additional caloric intake.

Identify the Sampling Method.

15. This is an example of stratified sampling. Perhaps the IRS auditor wishes to ensure that people above and below a certain income threshold were audited.

17. This is an example of stratified sampling. People in different age groups were selected so that the survey would not just sample the young or old.

19. This is an example of simple random sampling. It's easy to automate and carry out.

Type of Study.

21. Observational, not case-control.

23. Observational, not case-control.

25. Experiment; treatment group consists of plants treated with fertilizer; control group consists of plants treated with no fertilizer; neither placebo nor blinding are needed.

Which Type of Study?

27. An observational study would be best: this will preclude any deliberate under- or over-estimating which might arise.

29. An case-control observational study would be best: since the artificial flavoring is potentially harmful, it would not be ethical to require any study participants to take it.

31. An experimental study would be best: have some people take an aspirin a day, and a control group who take a placebo a day.

Experimental Results.

33. Since the proportion of people showing improvement in the treatment group was three times as high as in the control group, there is evidence the treatment is effective.

35. Since the proportion of people showing improvement in the treatment group was almost twice as high as in the control group, there is good evidence that the treatment is effective.

Margin of Error.

37. The confidence interval for the percentage of voters in favor of the Republican candidate is obtained by subtracting and adding the margin of error, namely 2.5%, from the sample statistic, which is 53%. This yields: 50.5% to 55.5%. Since this is (just) above the 50% mark, it is probably safe for the Republican party to plan a victory party.

39. The confidence interval for the January unemployment rate is obtained by subtracting and adding the margin of error of 0.2% from 4.2%, which yields 4.0% to 4.4%. The confidence interval for the February unemployment rate is obtained by subtracting and adding the margin of error of 0.2% from 4.3%, which yields 4.1% to 4.5%. Since these confidence intervals overlap so much, we cannot conclude that unemployment rates increased as claimed.

Interpreting Real Studies.

41. a. The goal was to determine what percentage of adults thought their children would have a higher standard of education than they had. Population is all parents; population parameter is percentage of all parents who think their children will have a higher standard of education than they had.
b. Sample is 748 parents surveyed; sample statistic is percentage of parents in sample who respond yes to the question.
c. The confidence interval is 63% plus or minus 3.6%, namely 59.4% to 66.6%.

43. a. The goal of the study was to determine the unemployment rate for September 1999. Population is presumably all Americans of working age; population parameter is percentage of working-age people who are unemployed.
b. Sample is working-aged people in the people 60,000 households surveyed; sample statistic is percentage of those people who are unemployed.
c. The confidence interval is 4.2% plus or minus 0.2%, namely 4.0% to 4.4%.

45. a. The goal of the study was to determine the percentage of adult Americans who believe that humans would be cloned within the next 50 years. Population is all adult Americans; population parameter is percentage of all adult Americans who believe that humans will be cloned within 50 years.

b. Sample is the 1546 adults Americans surveyed; sample statistic is percentage of those people who responded yes to the question.
c. The confidence interval is 51% plus or minus 3%, namely 48% to 54%.

Unit 5B Should You Believe a Statistical Study?

Bias.

1. There is possible conflict of interest. It's debatable if an unfavorable review would be aired on a network owned by Disney.

3. There is probably selection bias. People who vote during this period are unlikely to represent a good cross-section of all voters.

5. There is probably selection bias. Marines are unlikely to represent a good cross-section of all 18- to 24-year olds.

7. There is possible conflict of interest. It is in the interest of the company to find that their product poses no threat.

9. Solutions will vary.

Stat-Bytes.

11. What was the sample and sample size? How was "great deal of confidence" measured? Which military leaders were considered?

13. What was the sample and sample size? Is self-reporting accurate?

15. What was the sample for the first estimate? The sample size? Similar questions could be asked concerning the second estimate. Furthermore, since a ten-year period is implied, the revenue growth figure is hard to access without relevant inflation and cost of living increase information.

Accurate Headlines.

17. Apart from the fact that "98% of (all) movies" is not the same as "98% of top rental movies," the summary merely claimed that 98% of the top rental movies contained drug use, or drinking or smoking.

19. The headline could be true, but it cannot be justified by the cited evidence: the sample is extremely small and there could be conflict of interest bias in the study.

Unit 5C Statistical Tables and Graphs

Qualitative vs. Quantitative.

1. This variable is qualitative. Blood types are associated with names, e.g., Type O+, not numbers.

3. This variable is qualitative. Like grades, the responses are measuring quality.

5. This variable is qualitative. The responses are nonnumerical.

7. This variable is qualitative. The choices are nonnumerical.

Frequency Tables.

9.

Grade	Freq.	Rel. freq.	Cum. freq.
A	4	0.17	4%
B	7	0.29	11%
C	8	0.33	19%
D	3	0.13	22%
F	2	0.08	24%
Total	24	1.00	24%

Binned Frequency Tables.

11.

Bin	Freq.	Rel. freq.	Cum. freq.
95–99	2	0.08	2%
90–94	2	0.08	4%
85–89	6	0.24	10%
80–84	3	0.12	13%
75–79	4	0.16	17%
70–74	1	0.04	18%
65–69	4	0.16	22%
60–64	0	0.00	22%
55–59	1	0.04	23%
50–54	1	0.04	24%
<50	1	0.04	25%
Total	25	1.00	25%

Statistical Graphs.

13. a. The teams are qualitative categories.

b. Here is a bar graph for this data:

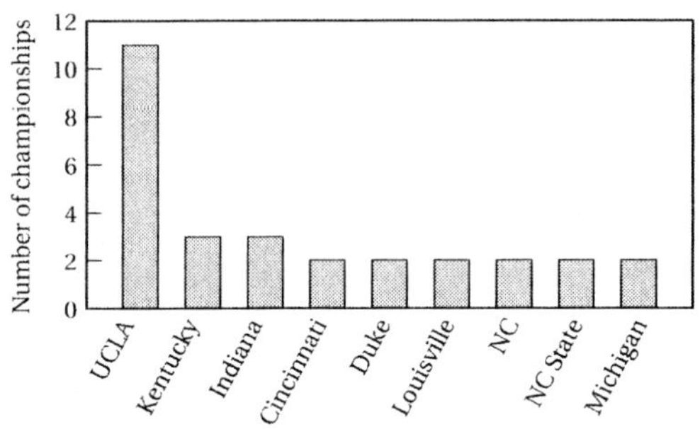

c. UCLA won about four times as many championships as any other team, and the other teams performances are all on a par with each other.

15. a. The ethnic groups are qualitative categories.

b. Here is a bar graph for this data:

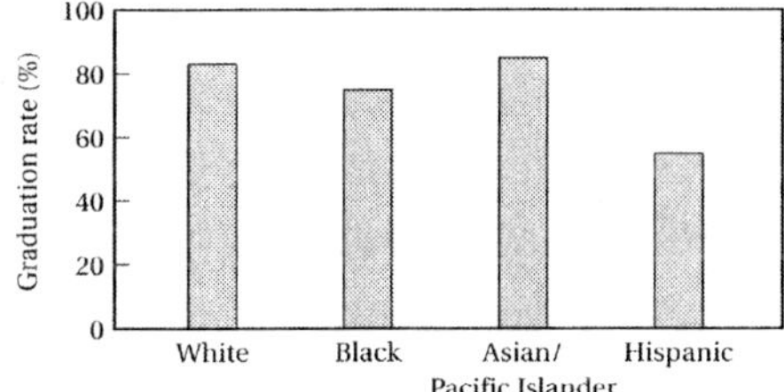

c. The first three graduation rates do not differ that much from each other; all are significantly higher than the rate for Hispanics.

17. a. The land masses are qualitative categories.

b. Here is a bar graph for this data:

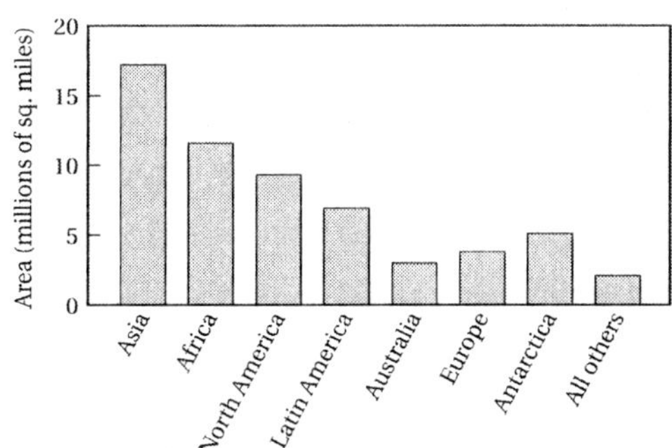

c. The land masses come in a great variety of areas, with no one size dominating.

19. a. The religions are qualitative categories.

b. Here is a bar graph for this data:

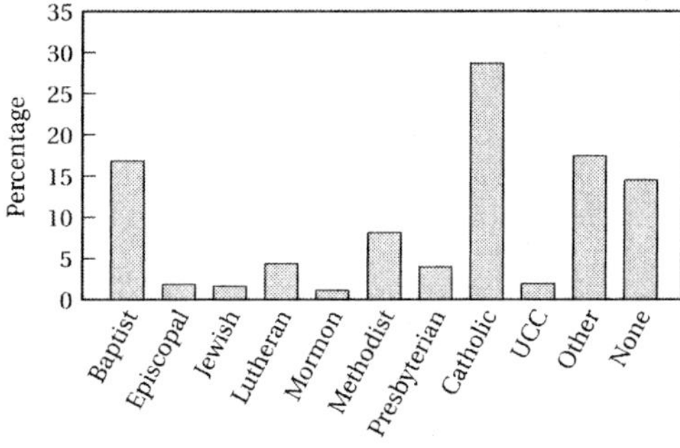

c. Only two religions, namely Catholics and Baptists, dominate, accounting for about 17% and 29% of the students respectively. No other religion even accounts for 10% of the students.

21. a. The years are quantitative categories.

b. Here is a line graph for the data:

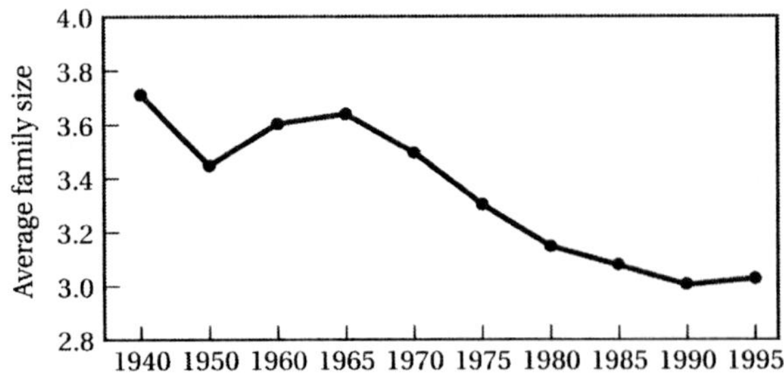

c. The average family size dropped from 1940 to 1950, almost regained its former level by 1965, and then declined steadily until 1990. There was a slight increase in the first half of the 1990s.

Unit 5D Graphics in the Media

1. Net Grain Production.

a. China, India, and Russia all had to import grain in 1990.

b. China, India, and Russia are all expected to need to import grain in 2030.

c. It suggests that considerably more grain will need to be grown around the world over the next 30 years.

3. Stack Plot.

a. The death rate for cancer increased; for tuberculosis it decreased; for cardiovascular disease it increased then decreased with overall decrease; and for pneumonia it decreased.

b. The death rate for cardiovascular disease reached a maximum of about 500 (deaths per 100,000) in 1950.

c. The death rate for cancer in 2000 was about 210 (deaths per 100,000).

d. If the lines of the graph are extended, it appears that cancer may overtake cardiovascular disease as the leading cause of death.

5. Federal Spending.

a. The percentage of the budget that went to net interest was: about $78\% - 70\% = 8\%$ in 1980; about $85\% - 70\% = 15\%$ in 1990; about $92\% - 77\% = 15\%$ in 1995; and about $82\% - 74\% = 12\%$ in 2000.

b. The percentage of the budget that went to defense was: about $78\% - 26\% = 52\%$ in 1960; about $70\% - 47\% = 23\%$ in 1980; and about $74\% - 57\% = 17\%$ in 2000.

c. The percentage of the budget that went to payment for individuals was: about 27% in 1960; about 47% in 1980; and about 57% in 2000.

7. School Segregation. There are regional differences. E.g., the probability that a black student would have white classmates is, in general, much higher in the northern states, and lower in many parts of the south and southwest, as well as in some specific cities (Chicago, Detroit, New York). Other southern cities stand out, such as Atlanta, which has a cluster of historically black colleges.

U.S. Age Distribution.

9. a. Each bar represents a percentage of the population for a particular age group in a particular year.

b. There has been a general decrease in the percentage of people in the youngest age category.

c. The data displayed here are truly three-dimensional.

Creating Graphics.

11. Percent Never Married. Two separate graphs work best here, here is the one for women:

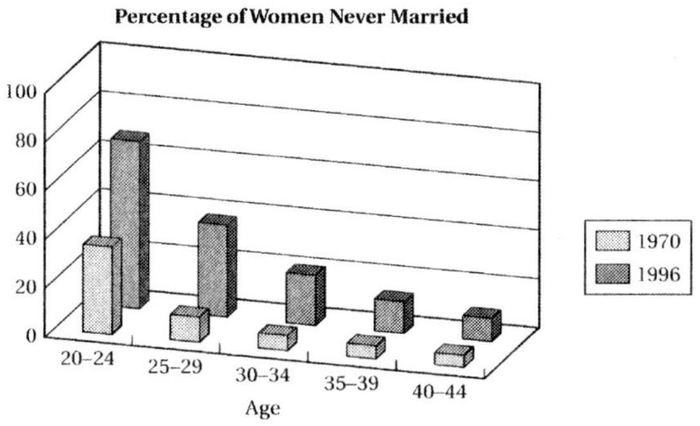

We see that women were marrying significantly older in the mid 90s than they were in 1970.

The following graph for men reveals that men were marrying quite a bit older in the mid 90s than they were in 1970.

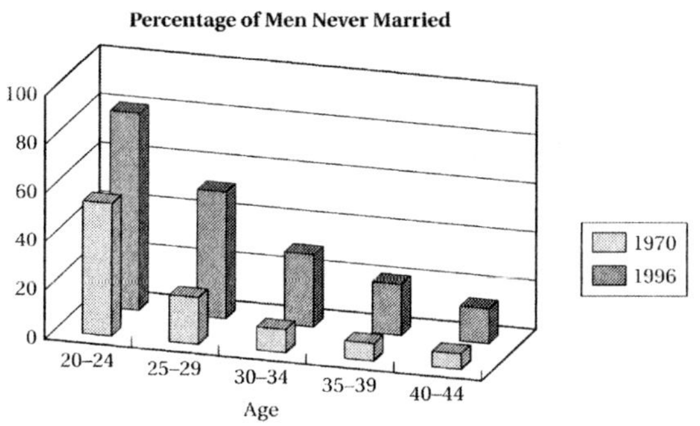

13. Daily Newspapers. A single graph can display this data:

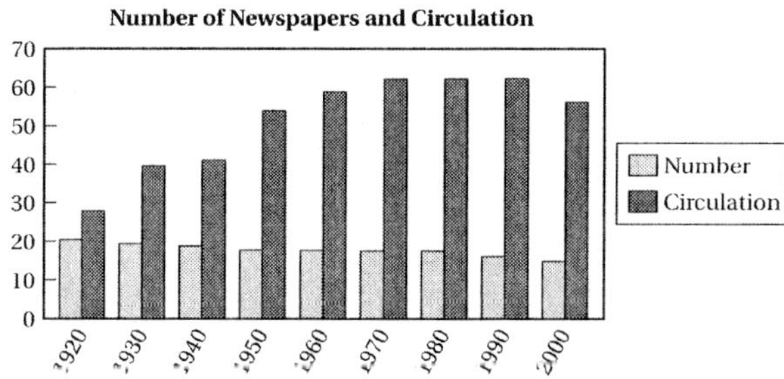

We see that while there has been a gradual drop in the number of daily papers over the decades in question, there was a dramatic increase in circulation (perhaps due to increased literacy rates?).

15. Volume Distortion. While the height of the 1997 TV in the figure is about 4 times the height of the 1980 TV, the visual perception is based on the area or volume discrepancy, depending on whether one pays attention to the depth of the TVs as depicted. The area of the 1997 TV is about $4 \times 4 = 16$ times the area of the 1980 TV. The volume of the 1997 TV is about $4 \times 4 \times 4 = 64$ times the area of the 1980 TV. Either way one interprets the picture, one gets the impression of a growth factor far in excess of 4.

17. Braking Distances. Since the zero point for braking distance is not on the graph, the graph is deceptive and a casual reader might come away with the mistaken impression that the Lincoln breaking distance is more than twice that for Oldsmobiles. In fact, the braking distance of about 206 feet for Lincolns is only 11% greater than the braking distance of about 185 feet for Oldsmobiles, since $\frac{206-185}{185} = 11\%$. A fairer way to draw the display is to include the zero point for braking distances.

19. Computer Sales. Here is a graph with an ordinary, unadjusted vertical scale:

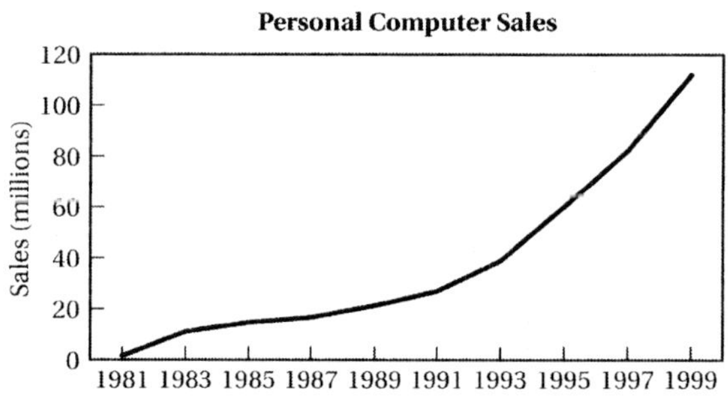

Here is one with an exponential scale:

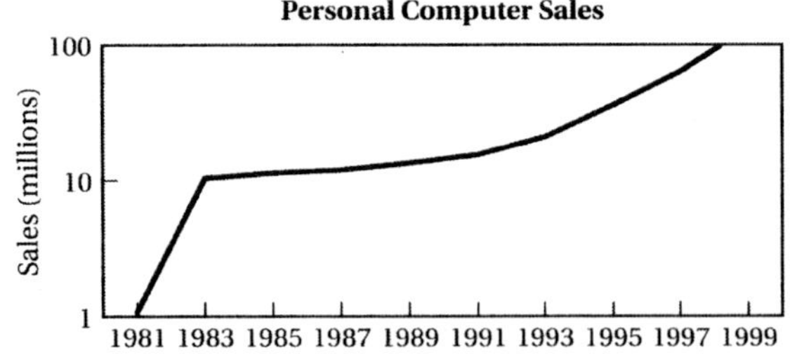

Unit 5E Correlation and Causality

Types of Correlation.

1. Body weight could be measured in pounds; calorie intake in calories. We'd expect a weak positive correlation: increased calorie intake should result in increased body weight in some people.

3. The rate of pedaling would be measured in revolutions per minute; the bicycle speed in miles per hour. We'd expect a strong positive correlation: the faster you pedal, the faster the bike would go in general.

5. The blood-alcohol level could be measured in percentage alcohol in the bloodstream; the reaction time in seconds. We'd expect a strong positive correlation: for most people, the higher the blood-alcohol level, the greater the reaction time.

7. The weight of a car could be measured in pounds; the price in dollars. We'd expect a weak positive correlation: the heavier the car, the more expensive it might be.

Making Scatter Plots.

9. a.

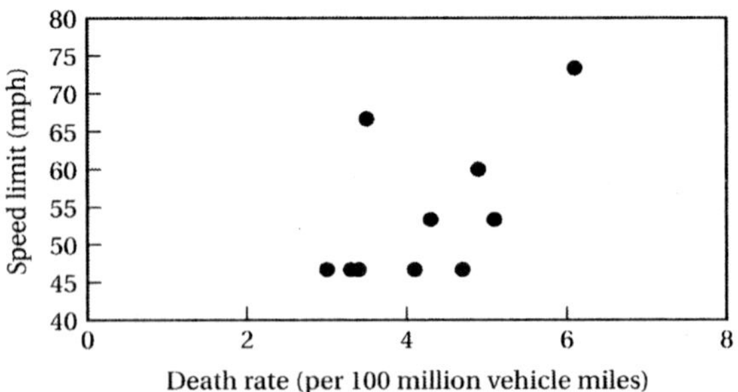

b. There is a weak positive correlation.

c. The correlation could be explained by asserting that in countries with higher speed limits, people tend to drive faster, and are more likely to have more frequent and more serious accidents.

11. a.

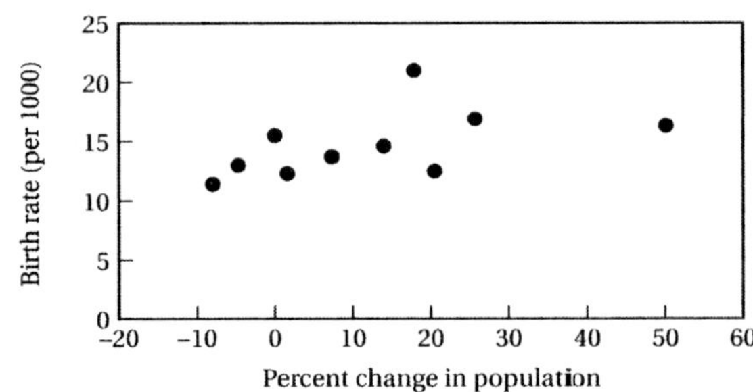

b. There is a weak positive correlation.

c. The correlation could be explained by asserting that in states that have experienced high growth rates, families tend to be larger.

13. a.

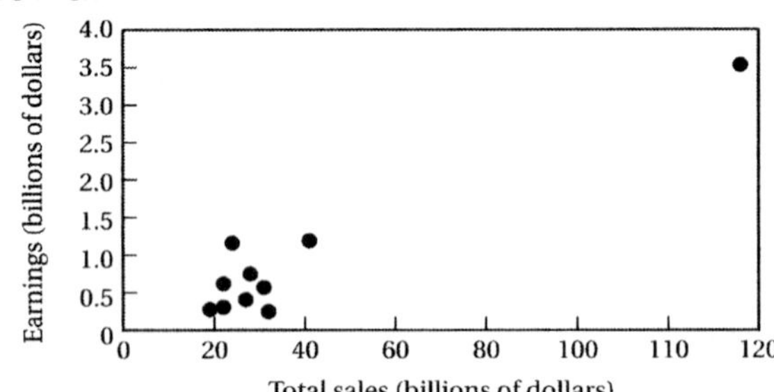

b. There is a strong positive correlation.

c. The correlation could be explained by asserting that the more retailers sell, the more profit they make.

Correlation and Causality.

15. There is a positive correlation between the crime rate and the number of people in prison. This may be due to a common underlying cause: increased criminal activity.

17. There is a positive correlation between gasoline prices and the number of airline passengers. This may be a case of direct cause: people opting to fly long distances rather than drive them.

19. The cost of deer hunting permits is positively correlated with deer sightings. This is probably a case of direct cause: the more expensive the hunting permit, the fewer people take one out and hunt, so that more live deer will be sighted.

21. Identifying Causes: Headaches.

a. Guideline 1.

b. Guideline 2.

c. Guideline 3.

d. The above observations suggest that perhaps your headaches are caused by bad air in the building in which you work (there is a "disease" called sick building syndrome).

23. Longetivity of Orchestra Conductors. People do not decide to become conductors until relatively later in life, say over 30 years of age. Having already survived this many years, a person's life expectancy is longer than average.

Unit 5F Characterizing a Data Distribution

Mean, Medians and Mode.

1. The mean is 671.37, obtained by dividing the sum of the six given numbers by 6. the median is 672.2, obtained by adding and dividing by 2 the two middle numbers when the numbers are listed in ascending order. The mode is 672.2, since it occurs more frequently than any of the other values.

3. The mean is 0.188, obtained by dividing the sum of the twelve given numbers by 12. the median is 0.165, obtained by adding and dividing by 2 the two middle numbers when the numbers are listed in ascending order. The mode is 0.16, since it occurs more frequently than any of the other values.

5. The mean is 9.5, obtained by dividing the sum of the fifteen given numbers by 15. the median is 10, since it is the middle value when the numbers are listed in ascending order. There are several modes: {5, 10, 15}, since these values occur more frequently than any of the others.

Appropriate Average.

7. The median would give a better description of the average income of all adults in a large city than the mean or mode. Using the mean would allow a small number of outliers at the high end (people with extremely high incomes) to pull the average up so that it would be unrepresentative of the data as a whole, and there is no real significance to the mode here.

9. The median would give a better description of the average number of times that people change jobs in their careers than the mean or mode. Using the mean would allow a small number of outliers at the high end (people who change jobs with great frequency) to pull the average up so that it was unrepresentative of the data as a whole, and there is no real significance to the mode here.

11. The mean or median would give a better description of the average daily high temperature for a month than the mode. If there are dramatic outliers at either the high or low end the median would be preferred, but there is no significance to the mode here.

13. Outlier Coke. The mean weight is 0.8124 pounds, obtained by dividing the sum of the seven given weights by 7. The median weight is 0.8161 pounds, which is the middle weight when all seven are arranged in ascending order.

The weight of 0.7901 pounds is an outlier. If we exclude this value, we are left with six weights whose mean is 0.8161 pounds and median is 0.8163 pounds.

Smooth Distributions.

15. This distribution has two peaks, and is thus considered to be bimodal. It is not symmetric, and it has high variation.

17. This distribution has one peak. It is symmetric, and it has moderate variation.

Describing Distributions.

19. Here we assume that A scores correspond to higher numerical scores, and so appear to the right of the C and B scores in the frequency distribution histogram.

a. We expect one peak, on the right, corresponding to the A scores.

b. This distribution would be left-skewed, since the mean scores would be less than the median score of an A grade.

c. The variation would be small, since two thirds of the scores were A's.

21. a. We expect two peaks, corresponding to two different classes of vehicles.

b. This distribution would be roughly symmetric, since about half of the vehicles are compact cars.

c. The variation would be large, since SUV's weigh a lot more than compact cars, so the spread of the data values is large.

23. a. We expect one peak, corresponding to a summer peak in sales.

b. This distribution would be roughly symmetric.

c. The variation would be moderate, since the climate supports swimming related activities the whole year round. so that the sales figures for the 12 months will be quite spread out.

25. a. We expect one peak, corresponding to the mode.

b. This distribution would be right-skewed, because of outliers at the high end (players earning very high salaries).

c. The variation would be large here, since there are huge outliers on the high side (8-10 times the median salary) with a lot of salaries lying between the high and the median. The salaries below the median are probably close to the median.

Unit 6A Fundamentals of Probability

Review of the Multiplication Principle.

1. a. There are 8 color selections to be made, and 3 style selections, so the total number of choices of car is $8 \times 3 = 24$.

b. There are 2 modes (AM or FM) for each button, and 5 buttons, so the total number of radio stations you can preset is $2 \times 5 = 10$.

c. There are 8 colors for each pattern, and 4 patterns, so the total number of wallpaper styles available is $8 \times 4 = 32$.

d. There are 6 TV sets, and 5 VCR players, so the total number of TV/VCR systems you can assemble is $6 \times 5 = 30$.

Theoretical Probabilities.

3. Each possible last digit represents an outcome, so there are 10 possible outcomes. It's reasonable to assume that these are equally likely outcomes. Only one of these—namely the digit 0—represents an event of interest, so the probability of meeting somebody with a phone number that ends in 0 is $\frac{1}{10} = 0.1$.

5. Each possibility {BB,BG,GB,GG} represents an outcome—e.g., BG denotes a boy followed by a girl—so there are 4 possible outcomes. If we assume that these are equally likely outcomes, then since only one of these—namely, GG—represents an event of interest, the probability of a randomly selected two-child family having two girls is $\frac{1}{4} = 0.25$.

7. There are 36 equally likely outcomes, namely $\{(1,1),(1,2),\ldots,(1,6),(2,1),(2,2),\ldots,(6,6)\}$, where, e.g., (2,1) denotes 2 on the first die and 1 on the second die. Since rolling a sum of 5 corresponds to one of the 4 outcomes $\{(1,4),(2,3),(3,2),(4,1)\}$, the probability of rolling a sum of 5 on a roll of two dice is $\frac{4}{36} = 0.111$.

9. There are 22 possible outcomes, since there are 22 pairs of socks. Assuming these are equally likely outcomes, 6 of these (the blue ones) are of interest, so the probability of selecting a pair of blue socks from the drawer is $\frac{6}{22} = 0.2727$.

Empirical Probabilities.

11. Based on the observation of 12 correct forecasts out of the last 30, the empirical probability of his next forecoast being correct is $\frac{12}{30} = 0.4$.

13. Based on the observation of hitting 86% of the past free throws, so far this year, the empirical probability of her next free throw being successful is 0.86.

15. Based on the 34 observed correct diagnoses out of 100, the empirical probability of his diagnosis being correct is $\frac{34}{100} = 0.34$.

Event *Not* Occurring.

17. The probability of rolling a 4 with a fair die is $\frac{1}{6}$, and so the probability of not rolling a 4 is $1 - \frac{1}{6} = \frac{5}{6} = 0.833$.

19. The probability that a 76% free shooter will be successful on his next free throw is 0.76, so the probability that he will miss his next free throw is $1 - 0.76 = 0.24$.

21. There are 36 equally likely outcomes, $\{(1,1),(1,2),\ldots,(1,6),(2,1),(2,2),\ldots,(6,6)\}$, and rolling a sum of 7 corresponds to one of the 6 outcomes $\{(1,6),(2,5),(3,4),(4,3),(5,2),$

(6,1)}, so that the probability of rolling a sum of 7 is $\frac{6}{36}$, and the probability of not rolling a sum of 7 is $1 - \frac{6}{36} = \frac{30}{36} = 0.833$.

23. The probability of rolling an odd number with a fair die is $\frac{3}{6}$, since 3 outcomes (namely 1, 3, or 5) out of 6 equally likely ones are of interest. The probability of not rolling an odd number is $1 - \frac{3}{6} = \frac{3}{6} = 0.5$.

25. Gender Politics. There are 100 possible outcomes, and since we meet a delegate at random, we will assume that these are equally likely outcomes. Since there are $28 + 16 + 4 = 48$ men present, the probability that the delegate will be a man is $\frac{48}{100}$, and hence the probability the delegate will not be a man is $1 - \frac{48}{100} = \frac{52}{100} = 0.52$. There are $21 + 28 = 49$ Republicans present, and so the probability that the delegate will be a Republican is $\frac{49}{100}$, and hence the probability the delegate will not be a Republican is $1 - \frac{49}{100} = \frac{51}{100} = 0.51$.

Probability Distributions.

27. Each possibility {HHHH,HHHT,HHTH, HHTT,HTHH,HTHT,HTTH,HTTT,THHH, THHT,THTH,THTT,TTHH,TTHT,TTTH, TTTT} represents an outcome—e.g., HHTH denotes two heads followed by a tail followed by a head—so there are 16 possible outcomes, which we assume are equally likely. Let's focus on the number of heads obtained, which can be 0, 1, 2, 3 or 4: P(0 heads) = P(TTTT) = $\frac{1}{16}$ = 0.0625; P(1 head) = P(HTTT,THTT,TTHT,TTTH) = $\frac{4}{16}$ = 0.25; P(2 heads) = P(HHTT,HTHT,HTTH, THHT,THTH,TTHH) = $\frac{6}{16}$ = 0.375; P(3 heads) = P(HHHT,HHTH,HTHH,THHH) = $\frac{4}{16}$ = 0.25; and finally P(4 heads) = P(HHHH) = $\frac{1}{16}$ = 0.0625.

29. Fair Coins? The coins appear to be unfair. There should be about 125 occurrences of 0 heads and 3 tails, since the probability of this is $\frac{1}{8} = 0.125$.

The Odds.

31. When rolling a fair die, the probability of A (the event of getting a 1 or a 2) is $\frac{2}{6} = \frac{1}{3}$, the probability of not A is $1 - \frac{1}{3} = \frac{2}{3}$, and so the odds of getting a 1 or a 2 are:

$$\text{Odds for A} = \frac{\text{P(A)}}{\text{P(not A)}} = \frac{\frac{1}{3}}{\frac{2}{3}} = \frac{1}{2},$$

or, 1 to 2.

33. When rolling a fair die, the probability of A (the event of getting a 5 or a 6) is $\frac{2}{6} = \frac{1}{3}$, the probability of not A is $1 - \frac{1}{3} = \frac{2}{3}$, and so the odds of getting a 5 or a 6 are:

$$\text{Odds for A} = \frac{\text{P(A)}}{\text{P(not A)}} = \frac{\frac{1}{3}}{\frac{2}{3}} = \frac{1}{2},$$

or, 1 to 2.

Gambling Odds.

35. Odds of 3 to 4 mean that for every $4 you bet, you will gain $3. Betting $20 is equivalent to making five $4 bets, so you will gain five times $3, namely $15.

37. Deceptive Odds. Since P(A) = 0.99, then P(not A)= 0.01, and

$$\text{Odds for A} = \frac{\text{P(A)}}{\text{P(not A)}} = \frac{0.99}{0.01} = \frac{99}{1},$$

or, 99 to 1. Meanwhile, since P(B) = 0.96, then P(not B)= 0.04, and

$$\text{Odds for B} = \frac{\text{P(B)}}{\text{P(not B)}} = \frac{0.96}{0.04} = \frac{24}{1},$$

or, 24 to 1. A small shift in probability (relatively speaking) lead to a large change (by a factor of 4) in the odds; alternatively, two events with radically different odds can have deceptively similar probabilities.

39. Solutions will vary.

Unit 6B Combining Probabilities

And Probabilities.

1. These events are independent. The probability that all three tickets are winners is the product of the probabilities that any one ticket is a winner, i.e., $\frac{1}{10} \times \frac{1}{10} \times \frac{1}{10} = 0.001$.

3. These events are dependent. The probability that all six people selected are women is the probability that the first one is a women, times the probability that the second one is a woman given that the first one is a woman, times the probability that the third one is a woman given that the first two are women, and so on, down to the probability that the sixth one is a woman given that the first five are women, i.e.,

$$\frac{12}{24} \times \frac{11}{23} \times \frac{10}{22} \times \frac{9}{21} \times \frac{8}{20} \times \frac{7}{19} = 0.00686.$$

5. These events are independent. The probability that both chips are defective is the product of the probabilities that each one is, i.e., $0.015 \times 0.015 = 0.000225$.

7. These events are independent. The probability that all we toss four tails followed by four heads is the probability that we toss four tails times the probability that we toss four heads, i.e.,

$$\frac{1}{16} \times \frac{1}{16} = 0.00391.$$

Either/Or Probabilities.

9. These events are non-overlapping, and the probability of rolling either a 1 or a 2 is the sum of the individual probabilities, i.e., $\frac{1}{6} + \frac{1}{6} = \frac{1}{3} = 0.333$.

11. These events are non-overlapping, and the probability of getting a sum of either 7 or 8 is the sum of the individual probabilities, which were found in Example 8 in Unit 6A in the text. We get $\frac{6}{36} + \frac{5}{36} = \frac{11}{36} = 0.306$.

13. These events are overlapping, and the probability of selecting either a woman or a Democrat is the sum of the individual probabilities minus the probability of selecting a female Democrat, i.e., $\frac{25+25}{100} + \frac{25+25}{100} - \frac{25}{100} = \frac{75}{100} = 0.75$.

15. These events are overlapping, and the probability of drawing either a king or a heart is the sum of the individual probabilities minus the probability of drawing both, i.e., $\frac{4}{52} + \frac{13}{52} - \frac{1}{52} = \frac{4}{13} = 0.308$.

At Least Once Problems.

17. The probability of getting at least one 6 in five rolls of a single die is 1 minus the probability of getting no 6s in five rolls, namely,

$$1 - \left(\frac{5}{6}\right)^5 = 0.598.$$

19. The probability of getting at least one head when tossing three coins is 1 minus the probability of getting no heads in three tosses, namely,

$$1 - \left(\frac{1}{2}\right)^3 = 0.875.$$

21. The probability of getting rain at least once in five days is 1 minus the probability of not getting rain on each of the five days, namely,

$$1 - (1 - 0.2)^5 = 0.672.$$

23. The probability of drawing at least one king in 20 draws is 1 minus the probability of not drawing a king on each of the 20 draws, namely,

$$1 - \left(1 - \frac{4}{52}\right)^{20} = 0.798.$$

25. Probability and Court.

a. These events are overlapping. The probability that a randomly selected

defendant either pled guilty or was sent to prison is the sum of the individual probabilities minus the probability that the person pled guilty and was sent to prison, i.e.,

$$\frac{392 + 564}{1028} + \frac{392 + 58}{1028} - \frac{392}{1028} = \frac{507}{514} = 0.986.$$

b. These events are overlapping. The probability that a randomly selected defendant either pled not guilty or was not sent to prison is the sum of the individual probabilities minus the probability that the person pled not guilty and was not sent to prison, i.e.,

$$\frac{58 + 14}{1028} + \frac{564 + 14}{1028} - \frac{14}{1028} = \frac{159}{257} = 0.619.$$

27. Polling Calls.

a. These events are dependent if the pollster makes a point to eliminate people already called, since the probability that the first person called is a Republican is $\frac{25}{45}$, and the probability that the second person called is also a Republican is $\frac{24}{44}$. If calls are made without regard to who has already been called, the events would be independent. The former is the more likely scenario.

b. Assuming dependence, the probability that the first two calls are to Republicans is

$$\frac{25}{45} \times \frac{24}{44} = 0.303$$

c. Assuming independence, the probability that the first two calls are to Republicans is

$$\frac{25}{45} \times \frac{25}{45} = 0.309$$

d. The answers are very close (this is because the list of names is fairly large, the difference gets less noticable for very long lists but is significant for very short ones).

29. Better Bet for the Chevalier. The probability of rolling at least one double-6 in 25 tries is

$$1 - \left(1 - \frac{1}{36}\right)^{25} = 0.506,$$

which is just over 50%. Hence, had he made his bet for 25 rather than 24 rolls, he would have had a slight edge, and over time he would have won money.

31. Miami Hurricanes.

a. Given the historical record, the empirical probability that Florida is hit by a hurricane next year is $\frac{1}{40} = 0.025$.

a. The probability that Florida is hit by a hurricane in two consecutive years is is $\frac{1}{40} \times \frac{1}{40} = 0.000625$.

c. The probability that Florida is hit by a hurricane at least once in the next ten years is

$$1 - \left(1 - \frac{1}{40}\right)^{10} = 0.224.$$

Unit 6C The Law of Large Numbers

1. Understanding the Law of Averages. You shouldn't expect to get exactly 5000 heads, no more than you expect to get exactly one head it you toss a fair coin twice. The law of averages tells us that the more often we toss, the closer the proportion of heads will approach 0.5, so we do expect to get approximately 5000 heads.

Expected Value in Games.

3. There is only 1 way, out of 8 equally likely outcomes, to toss three heads in three tosses of a fair coin, so the probability of tossing three heads in three tosses of a fair coin is $\frac{1}{8}$, and the probability of not tossing three heads is $\frac{7}{8}$. Hence, the expected value of the game is

$$(\$5 \times \frac{1}{8}) + (-\$1 \times \frac{7}{8}) = -\$0.25,$$

i.e., you expect to lose a quarter per game on average.

While the outcome of one game cannot be predicted, over 100 games, you should expect to lose.

5. Note that rolling two even numbers two fair dice means getting one of the 9 outcomes {(2,2), (2,4), (2,6), (4,2), (4,4), (4,6), (6,2), (6,4), (6,6)}, and so the probability of rolling two even numbers is $\frac{9}{36} = \frac{1}{4}$, and the probability of not rolling two even numbers is $\frac{3}{4}$. Hence, the expected value of the game is

$$(\$5 \times \frac{1}{4}) + (-\$1 \times \frac{3}{4}) = \$0.50,$$

i.e., you expect to win 50 cents per game on average.

While the outcome of one game cannot be predicted, over 100 games, you should expect to win.

Insurance Claims.

7. There are four events here, each with a probability and value to the company: a policy purchase, a \$20,000 claim, a \$50,000 claim and a \$100,000 claim. Thus, the expected value to the insurance company of each policy is

$$(\$1000 \times 1) + (-\$20,000 \times \frac{1}{100}) +$$

$$(-\$50,000 \times \frac{1}{200}) + (-\$100,000 \times \frac{1}{500})$$

$= \$350$. If the company sells 100,000 policies, its expected profit is $100,000 \times \$350 = \$35,000,000$.

9. Expected Wait. Since you arrive randomly, the actual wait time is between 0 and 30 mins, and is just as likely to be under 5 minutes as over 25 minutes. It seems reasonable to argue that the wait time has a symmetric distribution, with range 0 to 30, and mean (and median) equal to 15 minutes.

Powerball.

11. There are ten events here, each with a probability and value to you: buying a lottery ticket, winning the \$30 million jackpot, a \$100,000 win, a \$5000 win, two different \$100 wins, two different \$7 wins, a \$4 win and a \$3 win.

Thus, your expected win for each ticket purchased is

$$(-\$1 \times 1) + (\$30,000,000 \times \frac{1}{80,089,128}) +$$
$$(\$100,000 \times \frac{1}{1,953,393}) + (\$5000 \times \frac{1}{364,042}) +$$
$$(\$100 \times \frac{1}{8879}) + (\$100 \times \frac{1}{8466}) +$$
$$(\$7 \times \frac{1}{207}) + (\$7 \times \frac{1}{605}) + (\$4 \times \frac{1}{188}) +$$
$$+(\$3 \times \frac{1}{74}) = -\$0.4302,$$

i.e., you expect to lose 43 cents per ticket.

Over the course of a year, your expected loss is about $365 \times \$0.43 = \$157$.

Big Game.

13. There are ten events here, each with a probability and value to you: buying a lottery ticket, winning the $3 million jackpot, a $150,000 win, a $5000 win, a $150 win, a $100 win, two different $5 wins, a $2 win and a $1 win. Thus, your expected win for each ticket purchased is

$$(-\$1 \times 1) + (\$3,000,000 \times \frac{1}{76,275,360}) +$$
$$(\$150,000 \times \frac{1}{2,179,296}) + (\$5000 \times \frac{1}{339,002}) +$$
$$(\$150 \times \frac{1}{9686}) + (\$100 \times \frac{1}{7705}) +$$
$$(\$5 \times \frac{1}{220}) + (\$5 \times \frac{1}{538}) +$$
$$(\$2 \times \frac{1}{102}) + (\$1 \times \frac{1}{62}) = -\$0.7809,$$

i.e., you expect to lose 78 cents per ticket.

Over the course of a year, your expected loss is about $365 \times \$0.78 = \$285$.

15. Extra Points in Football. If the team opts for trying to score 1 extra point, there are two events: success (with probability 0.94) and failure (with probability 0.06). The expected number of points is

$$(1 \times 0.94) + (0 \times 0.06) = 0.94.$$

If the team opts for trying to score 2 extra points, there are two events: success (with probability 0.37) and failure (with probability 0.63). The expected number of points is

$$(2 \times 0.37) + (0 \times 0.63) = 0.74.$$

Based on these expected values, it makes more sense to go for the first option in most cases, since on average more extra points will be won. However, if the team were one point behind, and the game were nearly over, it might be worth going for 2 extra points, which could clinch victory.

17. Household Size. If we interpret the expected number of people in an American household to mean the expected value of the size of a randomly selected American household, then we can argue that according to the given categories, there are three possible events. The first is that the household selected has 1 or 2 occupants, with value 1.5 persons, which happens with probability 0.57, the second is that the household selected has 3 or 4 occupants, with value 5.5 persons, which happens with probability 0.32, and the third is that the household selected has 5 or more occupants, with value 6 persons, which happens with probability 0.11. Hence, the expected number of people in an American household is

$$(1.5 \times 0.57) + (3.5 \times 0.32) + (6 \times 0.11),$$

which comes out to 2.635.

19. Gambler's Fallacy and Dice.

a. Even numbers have come up 45 times out of 100, so you have won \$45 but lost $\$(100 - 45) = \$55$, so you are down $\$55 - \$45 = \$10$.

b. In total, even numbers have come up $45 + 47 = 92$ times out of 200, so you have won \$92 but lost $\$(200 - 92) = 108$, so you are down $\$108 - \$92 = \$16$.

c. By this stage, even numbers have come up $45 + 47 + 148 = 240$ times out of 500, so you have won \$240 but lost $\$(500 - 240) = \$260$, so you are down $\$260 - \$240 = \$20$.

d. If even numbers come up M times in the next 100 rolls, then overall they have come up $240 + M$ times in the first 600 rolls, and consequently odd numbers have have come up $600 - (240 + M) = 360 - M$ times. At that stage, you would have won $\$(240 + M)$, and had to pay $\$(360 - M)$. Breaking even means that the wins and losses cancel out, so we set $240 + M = 360 - M$, and solve for M, finding that $M = 60$. This is certainly possible, although not too likely!

e. The percentages of even numbers after 100, 200 and 500 rolls, respectively, were $\frac{45}{100} = 0.45 = 45\%$, $\frac{45+47}{200} = 0.46 = 46\%$ and $\frac{45+47+148}{500} = 0.48 = 48\%$, which are getting closer and closer to the predicted longterm 50% rate, despite your mounting losses.

21. Profitable Casino. If the casino can expect to earn \$0.07 per \$1 gambled, and \$100 million is wagered over the course of a year, them the total profit is $\$100,000,000 \times 0.07 = \$7,000,000$, i.e., \$7 million.

23. Mail Sweepstakes.

a. There are two possible outcomes here, a loss of 34 cents for you, or a \$1 million win. Your expected gain is:

$$(-\$0.34 \times 1) + (\$1,000,000 \times \frac{1}{10,000,000})$$

$= -\$0.24$, i.e., a 24 cent loss.

b. We assume that you subscribe to the three magazines one way or another, so again there are two outcomes, a loss of 34 cents plus the \$7 too much you pay for the subscriptions, or a \$1 million win. Your expected gain is:

$$(-\$7.34 \times 1) + ((\$1,000,000 - \$7) \times \frac{1}{10,000,000})$$

$= -\$7.24$, i.e., a \$7.24 cent loss.

Unit 6D Counting and Probability

Review of Factorials.

1. a. $7! = 7 \times 6 \times 5 \times 4 \times 3 \times 2 \times 1 = 5040.$

b. $\frac{7!}{3!} = \frac{7\times6\times5\times4\times3\times2\times1}{3\times2\times1} = 7 \times 6 \times 5 \times 4 = 840.$

c. $\frac{7!}{4!3!} = \frac{7\times6\times5\times4\times3\times2\times1}{(4\times3\times2\times1)(3\times2\times1)} = \frac{7\times6\times5}{3\times2\times1} = 35.$

d. $\frac{12!}{5!} = \frac{12\times11\times10\times9\times8\times7\times6\times5\times4\times3\times2\times1}{5\times4\times3\times2\times1} = 12 \times 11 \times 10 \times 9 \times 8 \times 7 \times 6 = 3,991,680.$

e. $\frac{12!}{8!(12-8)!} = \frac{12\times11\times10\times9\times8\times7\times6\times5\times4\times3\times2\times1}{(8\times7\times6\times5\times4\times3\times2\times1)(4\times3\times2\times1)} = \frac{12\times11\times10\times9}{4\times3\times2\times1} = 495.$

f. $\frac{21!}{20!} = \frac{21\times20!}{20!} = 21.$

Counting Possibilities.

3. This is arrangement with repetition, and so we have $10^4 = 10,000$ different four-digit addresses using the 10 available digits.

5. This is arrangement with repetition, and so we have $26^5 = 11,881,376$ different five-letter words.

7. This is arrangement without repetition, i.e., permutations, and so we have $6 \times 5 \times 4 \times 3 \times 2 = 720$ different five-letter passwords using the available 6 letters.

9. This is arrangement with repetition, and so we have $4^3 = 64$ different three-letter words using the available four letters.

11. This is combinations, as we are only interested in the makeup of the subcommittee, order plays no role. So we have ${}_{10}C_5 = 252$, different five-person subcommittees, using the ten available people.

13. This is combinations, as we are only interested in the makeup of the selection of compact disks taken, order plays no role. So we have ${}_{10}C_6 = 210$, different six-disk selections, using the ten available disks.

15. This is combinations, as we are only interested in the makeup of the subcommittee, order plays no role. So we have ${}_{12}C_4 = 495$ different five-person subcommittees, using the ten available people.

17. This is two arrangements with repetition—one for letters and another for numerals. The letters can be arranged in $26^3 = 17,576$ ways, and the numerals in $10^3 = 1000$ ways. By the multiplication principle, we have $17,576 \times 1000 = 17,576,000$ different license plates.

19. Ice Cream Shop.

a. By the multiplication principle, we have $12 \times 6 = 72$ different sundaes.

b. This is arrangement with repetition, since we specify which flavor goes in which of three positions and the same flavor can be used more than once, and so we have $12^3 = 1728$ different triple cones.

c. This is arrangement without repetition, i.e., a permutation, since we specify which flavor goes in which of three positions and the same flavor may not be used more than once, and so we have $12 \times 11 \times 10 = 1320$ different triple cones this time.

d. This is combinations, as we are only interested in the makeup of the cone, order plays no role. So we have ${}_{12}C_3$, which comes out to 220, different triple cones, using the twelve available flavors.

21. Pizza Hype. This is combinations, as we are only interested in the makeup of the pizza, order plays no role. If Luigi uses N different pizza toppings, then we have ${}_NC_3 = 84$ different three-topping pizzas. We wish to know N: some trial and error leads to the discovery that $N = 9$. Similarly, if Ramona uses M different pizza toppings, then we have ${}_MC_2 = 45$ different two-topping pizzas. We wish to know M: some trial and error leads to the discovery that $M = 10$.

Counting and Probability.

23. There are ${}_{32}C_5 = 201,376$ ways to select 5 different numbers from the 32 available, and only 1 of these matches a particular randomly selected set. So the desired probability is $\frac{1}{201,376}$.

25. There are ${}_{52}C_5 = 2,598,960$ ways to select 5 cards from the 52 available, and only 4 of these hands are desirable (those consisting of the ace, 2, 3, 4 and 5 of each suit). So the desired probability is $\frac{4}{2,598,960} = \frac{1}{649,740}$.

27. There are ${}_{10}P_3 = 10 \times 9 \times 8 = 720$ ways to guess the top 3 winners (in order) from the 10 available, and only 1 of these guessed matches the actual list of top 3 winners, so the desired probability is $\frac{1}{720}$.

29. If you are one of 10 people at the party, then the probability that a particular other person present does not have your birthday is $\frac{364}{365}$. Hence, the probability that at least one other person has your birthday is 1 minus the probability that none do, namely:

$$1 - \left(\frac{364}{365}\right)^9 = 1 - 0.9756 = 0.0244.$$

The probability that at least one pair of guests share the same birthday—but not necessarily yours—is 1 minus the probability that none do, namely:

$$1 - \left(\frac{364 \times 363 \times 362 \ldots \times 356}{365^9}\right)$$
$$= 1 - 0.8831 = 0.1169.$$

So while there is only roughly a 2.4% chance that somebody else has the same birthday as you, there is almost a 12% chance that some two (or more) people present share a birthday.

31. Hot Streaks.

a. The probability of winning five games in a row is $(0.48)^5$, or about $0.025 = 2.5\%$ (a 1 in 40 chance).

b. The probability of winning ten games in a row is $(0.48)^{10}$, or about $0.00065 = 0.065\%$ (a 1 in 1540 chance).

c. If an individual's chances of having a five-game hot streak is 2.5%, then out of 2000 players, we could expect $2.5\% \times 2000 = 50$ people to have a five-game hot streak.

d. If an individual's chances of having a ten-game hot streak is 0.065%, then out of 2000 players, we could expect $0.065\% \times 2000 = 1.3$ people to have a ten-game hot streak.

33. Joe DeMaggio's Record.

a. The probability that a player who gets a hit 40% of the time (and hence misses 60% of the time) will get at least one hit out of four tries is 1 minus the probability that he misses all four, i.e.,

$$1 - (0.6)^4 = 1 - 0.1296 = 0.8704,$$

namely an 87% chance.

b. The probability that a player who gets a hit 40% of the time will get a hit in 56 consecutive games—each game consisting of four at-bats—is therefore $0.8704^{56} = 0.0004$, or about 1 in 2500.

c. We need to repeat the computations in (a) and (b) above using a 0.300 batting average instead of 0.400. The probability that a player who gets a hit 30% of the time (and hence misses 70% of the time) will get at least one hit out of four tries is 1 minus the probability that he misses all four, i.e.,

$$1 - (0.7)^4 = 1 - 0.2401 = 0.7599,$$

namely a 76% chance.

The probability that such a player will get a hit in 56 consecutive games is thus $0.7599^{56} = 0.0000002$, or about a 1 in 5 million chance.

Unit 7A Growth: Linear versus Exponential

Linear or Exponential?

1. This is linear growth; after three years the population would be $15,000+(3\times 505) = 16,515$ people.

3. This is exponential growth. After one month your food bill would have been higher by $0.30 \times \$100 = \$30$, making it \$130; and after a second month it would have risen an additional $0.30\times\$130 = \$39$, making it \$169.

5. This is exponential decay. After one year the cost of a memory chip will drop $0.12 \times \$80 = \$9.60$, making the cost \$70.40; after a second year the cost will drop a further $0.12 \times \$70.40 = \$8.45$, to \$61.95.

7. This is linear growth; in three years your house will be worth $\$100,000 + (3 \times \$10,000) = \$130,000$.

Chessboard Parable.

9. Square 20 of the chessboard should hold $2^{19} = 524,288$ grains of wheat. At this point, there would be $2^{20} - 1 = 1,048,575$ grains of wheat on the board, weighing

$$1,048,575 \text{ grains} \times \frac{1 \text{ lb}}{7000 \text{ grains}},$$

or, about 150 lbs.

11. When the chessboard is full, there would be $2^{64} - 1 = 1.845 \times 10^{19}$ grains of wheat on it, weighing

$$1.845 \times 10^{19} \text{ grains} \times \frac{1 \text{ lb}}{7000 \text{ grains}}$$
$$= 2.636 \times 10^{15} \text{ lbs},$$

or, dividing by 2000, about 1.3×10^{12} tons.

Magic Money Parable.

13. After 15 days, you would have $2^{15} \times \$0.01 = \$327.68$.

15. A billion dollars is 100 billion cents, so we need to find the smallest N such that $2^N > 100,000,000,000$. Trying some values we find that $N = 10$ yields $2^N = 1024$; $N = 20$ yields $2^N = 1,048,576$; $N = 30$ yields $2^N = 1,073,741,824$; and $N = 40$ yields $2^N = 1,099,511,627,776$, So 30 is too small, and 40 works but may well be too large. Try $N = 35$, which yields $34,359,738,368$, which is too small. Next, $N = 36$ yields $68,719,476,736$, which is still too small, but $N = 37$ yields $137,438,953,472$. Hence, 37 days would elapse before you had your first billion dollars.

Bacteria in a Bottle.

17. At 11:50, which is 50 minutes after the bacteria started dividing, there are $2^{50} = 1.1259 \times 10^{15}$ bacteria in the bottle. Since the bottle is $\frac{1}{2}$ full 1 minute before 12:00, $\frac{1}{4}$ full 2 minutes before 12:00, $\frac{1}{8}$ full 3 minutes before 12:00, and so on, it must be $\frac{1}{2^{10}} = \frac{1}{1024}$ full 10 minutes before 12:00.

19. Knee-Deep in Bacteria. After two hours, the volume of the colony would be 1.3×10^{15} cubic meters (see text). To get the depth of the resulting layer, distributed evenly on the earth's surface, divide by the surface area of the earth, which was given as 5.1×10^{14} square meters. This yields

$$\frac{1.3 \times 10^{15} \text{ m}^3}{5.1 \times 10^{14} \text{ m}^2} = \frac{13 \times 10^{14} \text{ m}^3}{5.1 \times 10^{14} \text{ m}^2} = 2.55 \text{ m},$$

which is about 8.36 feet. This is a lot more than knee deep!

21. Human Doubling.

a.

Year	Population	Year	Population
2000	6.000×10^9	2550	1.229×10^{13}
2050	1.200×10^{10}	2600	2.458×10^{13}
2100	2.400×10^{10}	2650	4.915×10^{13}
2150	4.800×10^{10}	2700	9.830×10^{13}
2200	9.600×10^{10}	2750	1.966×10^{14}
2250	1.920×10^{11}	2800	3.932×10^{14}
2300	3.840×10^{11}	2850	7.864×10^{14}
2350	7.680×10^{11}	2900	1.573×10^{15}
2400	1.536×10^{12}	2950	3.146×10^{15}
2450	3.072×10^{12}	3000	6.291×10^{15}
2500	6.144×10^{12}		(Hi mom!)

b. If each person occupied 1 square meter of the 5.1×10^{14} square meters of the earth's surface area, then we would have 5.1×10^{14} people. From the table, this would happen sometime between 2800 and 2850. (Actually, it would happen sooner, since two thirds of the earth's surface is covered in water!)

c. If each person needs 10^4 square meters of the earth's surface to survive, then the number of people the earth could support can be found using unit analysis:

$$\frac{5.1 \times 10^{14}\ \text{m}^2}{10^4\ \text{m}^2/\text{person}} = 5.1 \times 10^{10}\ \text{people},$$

which is just over 50 billion people. From the table, this would happen soon after 2150, which is less than 150 years away!

d. If we had five times as much surface area at our disposal, then from (b) above, we would have enough room for $5 \times 5.1 \times 10^{14} = 2.51 \times 10^{15}$ people. This is smaller than the projected population for the year 3000, so we would *not* have enough room (never mind food!) for all of humanity in the solar system at that point.

Unit 7B Doubling Time and Half-Life

Logarithms.

1. a. False; 0.928 is between 0 and 1, so $10^{0.928}$ is between $10^0 = 1$ and $10^1 = 10$.

b. True; 3.334 is between 3 and 4, so $10^{3.334}$ is between $10^3 = 1000$ and $10^4 = 10,000$.

c. False; -5.2 is between -6 and -5, so $10^{-5.2}$ is between $10^{-6} = 0.000001$ and $10^{-5} = 0.0001$.

d. True; -2.67 is between -3 and -2, so $10^{-2.67}$ is between $10^{-3} = 0.001$ and $10^{-2} = 0.01$.

e. True; π is between 1 and 10, so $\log_{10}(\pi)$ is between $\log_{10}(1) = 0$ and $\log_{10}(10) = 1$.

3. a. Since $4 = 2^2$, we have $\log_{10}(4) = \log_{10}(2^2) = 2 \times \log_{10}(2) = 2 \times 0.301 = -0.602$.

b. Since $20,000 = 2 \times 10^4$, we have $\log_{10}(20,000) = \log_{10}(2 \times 10^4) = \log_{10}(2) + \log_{10}(10^4) = 0.301 + 4 = 4.301$.

c. Since $0.5 = 2^{-1}$, we have $\log_{10}(0.5) = \log_{10}(2^{-1}) = -1 \times \log_{10}(2) = -1 \times 0.301 = -0.301$.

d. Since $32 = 2^5$, we have $\log_{10}(32) = \log_{10}(2^5) = 5 \times \log_{10}(2) = 5 \times 0.301 = 1.505$.

e. Since $0.25 = 2^{-2}$, we have $\log_{10}(0.25) = \log_{10}(2^{-2}) = -2 \times \log_{10}(2) = -2 \times 0.301 = -0.602$.

f. Since $0.2 = 2 \times 10^{-1}$, we have $\log_{10}(0.2) = \log_{10}(2 \times 10^{-1}) = \log_{10}(2) + \log_{10}(10^{-1}) = 0.301 + (-1) = -0.699$.

Doubling Time.

5. Since the population increases by a factor of 2 in 3 hours, and 24 hours represents

$\frac{24}{3} = 8$ doubling periods, it will increase by a factor of $2^8 = 256$ in 24 hours.

Since 1 week represents $\frac{7\times24}{3} = 56$ doubling periods, the population will increase by a factor of $2^{56} = 7.2 \times 10^{16}$ in 1 week.

7. Since the population increases by a factor of 2 in 22 years, it will quadruple ($2^2 = 4$) in 2 doubling periods, i.e., in 44 years.

9. Since the population increases by a factor of 2 in 20 years, and 12 years represents $\frac{12}{20} = 0.6$ doubling periods, it will increase by a factor of $2^{0.6} = 1.51572$ in 12 years. Thus the population will grow from 15,600 to $15,600 \times 1.51572 = 23,645$ people in 12 years.

Since 24 years represents $\frac{24}{20} = 1.2$ doubling periods, it will increase by a factor of $2^{1.2} = 2.29740$ in 24 years. Thus the population will grow from 15,600 to $15,600 \times 2.29740 = 35,839$ people in 24 years.

11. Since the number of cells increases by a factor of 2 in 1.5 months, and 3 years represents $\frac{3\times12}{1.5} = 24$ doubling periods, the number of cells will increase by a factor of $2^{24} = 16,777,216$ in 3 years. Thus, 1 cell grows to almost 17 million cells in 3 years.

Since 4 years represents $\frac{4\times12}{1.5} = 32$ doubling periods, the number of cells will increase by a factor of $2^{32} = 4,294,967,296$ in 4 years. Thus, 1 cell grows to over 4 billion cells in 4 years.

World Population.

13. Assuming the world population increases by a factor of 2 in 45 years, then since 2010 is 10 years after 2000, when the population was 6 billion people, and 10 years represents $\frac{10}{45} = 0.222$ doubling periods, the world population will increase by a factor of $2^{0.222} = 1.167$ to 6 billion $\times$ 1.167 = 7 billion people by 2010.

Next, 2060 is 60 years after 2000, which is $\frac{60}{45} = 1.333$ doubling periods, so the world population will increase by a factor of $2^{1.333} = 2.520$ to 6 billion $\times$ 2.520 = 15.12 billion people by 2060.

Finally, 2100 is 100 years after 2000, which is $\frac{100}{45} = 2.222$ doubling periods, so the world population will increase by a factor of $2^{2.222} = 4.665$ to 6 billion$\times$4.665 = 28 billion people by 2100.

15. Rabbits.

M	Population at start of month	Population at end of month
1	$100 \times (1.07)^0 = 100$	$100 \times (1.07)^1 = 107$
2	$100 \times (1.07)^1 = 107$	$100 \times (1.07)^2 = 114$
3	$100 \times (1.07)^2 = 114$	$100 \times (1.07)^3 = 123$
4	$100 \times (1.07)^3 = 123$	$100 \times (1.07)^4 = 131$
5	$100 \times (1.07)^4 = 131$	$100 \times (1.07)^5 = 140$
6	$100 \times (1.07)^5 = 140$	$100 \times (1.07)^6 = 150$
7	$100 \times (1.07)^6 = 150$	$100 \times (1.07)^7 = 161$
8	$100 \times (1.07)^7 = 161$	$100 \times (1.07)^8 = 172$
9	$100 \times (1.07)^8 = 172$	$100 \times (1.07)^9 = 184$
10	$100 \times (1.07)^9 = 184$	$100 \times (1.07)^{10} = 197$
11	$100 \times (1.07)^{10} = 197$	$100 \times (1.07)^{11} = 210$
12	$100 \times (1.07)^{11} = 210$	$100 \times (1.07)^{12} = 225$
13	$100 \times (1.07)^{12} = 225$	$100 \times (1.07)^{13} = 241$
14	$100 \times (1.07)^{13} = 241$	$100 \times (1.07)^{14} = 258$
15	$100 \times (1.07)^{14} = 258$	$100 \times (1.07)^{15} = 276$

From the table, the doubling time is about 10 months, which agrees with the doubling time formula prediction of $\frac{70}{7} = 10$ months.

Doubling Time Formula.

17. The doubling time formula prediction for a 7% per year increase rate is $\frac{70}{7} = 10$ years; since the rate is less than 15%, the formula is valid. In 3 years, the CPI will increase by a factor of $1.07^3 = 1.225$.

19. The doubling time formula prediction for a 0.7% per month increase rate is $\frac{70}{0.7} =$ 100 months (i.e., 8 years and 4 months); since the rate is less than 15%, the formula is valid. In 1 year, prices will increase by a factor of $1.007^{12} = 1.087$. In 8 years, which is 96 months, they will increase by a factor of $1.007^{96} = 1.954$.

Half-Life.

21. Since 70 years is 2 half-lives, the amount of radioactive substance decreases by a factor of $(1/2)^2 = 1/4 = 0.25$, i.e., the amount present after 70 years is 0.25 times the amount we started with. Since 140 years is 4 half-lives, the amount of radioactive substance decreases by a factor of $(1/2)^4 = 1/16 = 0.0625$, i.e., the amount present after 140 years is 0.0625 times the amount we started with.

23. Since 24 hours is 2 half-lives, the amount of the drug in the bloodstream decreases by a factor of $(1/2)^2 = 0.25$, i.e., the amount present after 24 hours is 0.25 times the amount we started with. Since 36 hours is 3 half-lives, the amount of the drug decreases by a factor of $(1/2)^3 = 0.125$, i.e., the amount present after 36 hours is 0.125 times the amount we started with.

25. Since 40 years is 4 half-lives, the population decreases by a factor of $(1/2)^4 = 0.0625$, so after 40 years it will be 1 million $\times$ $0.0625 = 62,500$ animals. Since 70 years is 7 half-lives, it decreases by a factor of $(1/2)^7 = 0.0078125$, so after 40 years it will be 1 million $\times$ 0.0078125, i.e., 7813 animals.

27. Since 100 days is $\frac{100}{77} = 1.2987$ half-lives, the amount of Cobalt-56 decreases by a factor of $(1/2)^{1.2987} = 0.40649$, so the 1 kg will be reduced to 0.41 kg. Since 1 year is $\frac{365}{77} =$ 4.74026 half-lives, the amount of Cobalt-56 decreases by a factor of $(1/2)^{4.74026} =$ 0.03741, so the 1 kg will be reduced to about 0.04 kg.

29. Plutonium on Earth. Since 4.6 billion years is $\frac{4,600,000,000}{24,000} = 1.92 \times 10^5$ half-lives, the amount of Pu-2396 would have decreased by a factor of $(1/2)^{1.92\times10^5}$, which is close to zero, even when multiplied by 10 trillion tons! There would be so little left that no traces of it would be detectable today.

Half-Life Formula.

31. The approximate half-life formula prediction for a 10% per year decay rate is $\frac{70}{10} = 7$ years; since the rate is less than 15%, the formula is valid. In 100 years, the fraction of the forest remaining will be $(1/2)^{\frac{100}{7}} = 0.00005$, or about 1/20,000.

33. The approximate half-life formula prediction for a 7% per year decay rate is $\frac{70}{7} =$ 10 years; since the rate is less than 15%, the formula is valid. After 50 years, which is 5 half-lives, the elephant population will be reduced to $(1/2)^5 = 0.03125$ of its starting population. Multiplying by 10,000 we find that there would be about 313 elephants left.

Exact Formulas.

35. The approximate doubling time formula prediction for a 10% per year growth rate is $\frac{70}{10} = 7$ years. Using this, after 3 years, prices will be $2^{\frac{3}{7}} = 1.34590$ times their current level, so what costs \$1000 today should cost \$1345.90 in 3 years.

The exact doubling time formula for a 10% per year growth rate is

$$\frac{\log_{10}(2)}{\log_{10}(1+0.10)} = 7.2725 \text{ years.}$$

Using this exact value, after 3 years, prices will be $2^{\frac{3}{7.2725}} = 1.33100$ times their current level, so what costs \$1000 today should cost \$1331 in 3 years. Since the rate is less than 15%, the approximation formula is valid; as seen above, it gave fairly accurate results.

37. The approximate doubling time formula prediction for a 3% per year growth rate is $\frac{70}{3} = 23.33333$ years. Using this, after 50 years, the population will be $2^{\frac{50}{23.33333}} = 4.41636$ times its current level, so a nation of 100 million people today should have a population of 442 million in 50 years.

The exact doubling time formula for a 3% per year growth rate is

$$\frac{\log_{10}(2)}{\log_{10}(1+0.03)} = 23.44977 \text{ years.}$$

Using this exact value, after 50 years, the population would be $2^{\frac{50}{23.44977}} = 4.38391$ times its current level, 100 million people grow to 438 million people in 50 years. Since the rate is less than 15%, the approximation formula is valid; as seen above, it gave fairly accurate results.

Unit 7C Exponential Modeling

Review of Logarithms.

1. a. Since $2^x = 23$, take logarithms to get $\log_{10}(2^x) = \log_{10}(23)$, i.e., $x\log_{10}(2) = \log_{10}(23)$. Dividing by $\log_{10}(2)$ we get

$$x = \frac{\log_{10}(23)}{\log_{10}(2)} = 4.5236.$$

b. Since $10^x = 43$, take logarithms to get $\log_{10}(10^x) = \log_{10}(43)$, i.e., $x\log_{10}(10) = x = \log_{10}(43) = 1.6335$.

c. Since $3^{2x} = 12$, take logarithms to get $\log_{10}(3^{2x}) = \log_{10}(12)$, i.e., $2x\log_{10}(3) = \log_{10}(12)$. Dividing by $2\log_{10}(3)$ we get

$$x = \frac{\log_{10}(12)}{2\log_{10}(3)} = 1.1309.$$

d. Since $2 \times 8^x = 24$, we first divide by 2 so that $8^x = 12$, then take logarithms to get $\log_{10}(8^x) = \log_{10}(12)$, i.e., $x\log_{10}(8) = \log_{10}(12)$. Dividing by $\log_{10}(8)$ we get

$$x = \frac{\log_{10}(12)}{\log_{10}(8)} = 1.1950.$$

3. a. Since $\log_{10}(x) = 1.5$, we have,

$$10^{\log_{10}(x)} = 10^{1.5},$$

i.e., $x = 10^{1.5} = 31.6228$.

b. Since $\log_{10}(x) = -2$, we have,

$$10^{\log_{10}(x)} = 10^{-2},$$

i.e., $x = 10^{-2} = 0.01$.

c. Since $\log_{10}(x) = -1.5$, we have,

$$10^{\log_{10}(x)} = 10^{-1.5},$$

i.e., $x = 10^{-1.5} = 0.03162$.

d. Since $\log_{10}(2x) = 4$, we have,

$$10^{\log_{10}(2x)} = 10^4,$$

i.e., $2x = 10,000$, and dividing by 2 we get $x = 5000$.

Exponential Growth and Decay Laws.

5. a. $Q = 85,000 \times 1.024^t$, where Q is the population of the town, and t is the time (in years) since 1998.

b.

Year	Population
1	$85,000 \times 1.024^1 = 87,040$
2	$85,000 \times 1.024^2 = 89,129$
3	$85,000 \times 1.024^3 = 91,268$
4	$85,000 \times 1.024^4 = 93,458$
5	$85,000 \times 1.024^5 = 95,701$
6	$85,000 \times 1.024^6 = 97,998$
7	$85,000 \times 1.024^7 = 100,350$
8	$85,000 \times 1.024^8 = 102,759$
9	$85,000 \times 1.024^9 = 105,225$
10	$85,000 \times 1.024^{10} = 107,750$

c.

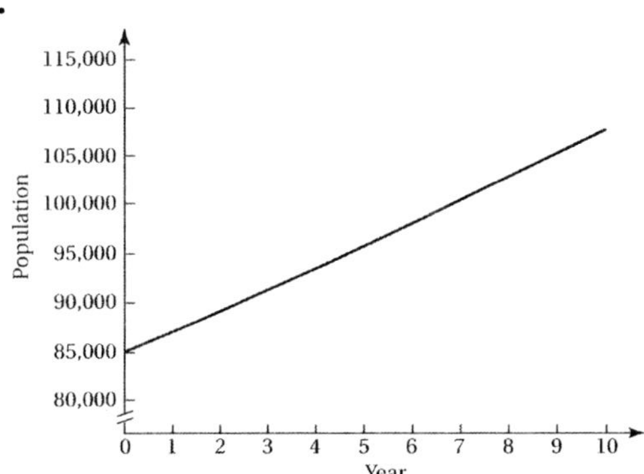

7. a. $Q = 1,000,000 \times 0.93^t$, where Q is the number of acres of old growth, and t is the time (in years) since the forest had 1 million acres.

b.

Year	Acres
1	$1,000,000 \times 0.93^1 = 930,000$
2	$1,000,000 \times 0.93^2 = 864,900$
3	$1,000,000 \times 0.93^3 = 804,357$
4	$1,000,000 \times 0.93^4 = 748,052$
5	$1,000,000 \times 0.93^5 = 695,688$
6	$1,000,000 \times 0.93^6 = 646,990$
7	$1,000,000 \times 0.93^7 = 601,701$
8	$1,000,000 \times 0.93^8 = 559,582$
9	$1,000,000 \times 0.93^9 = 520,411$
10	$1,000,000 \times 0.93^{10} = 483,982$

c.

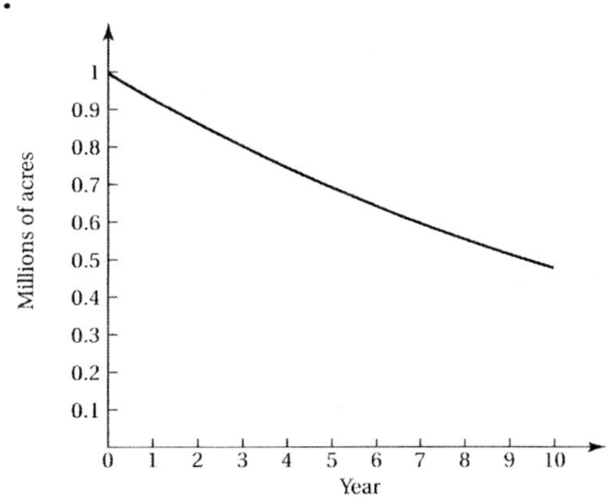

9. a. $Q = 125,000 \times 1.07^t$, where Q is the average price of a home, and t is the time (in years) since 1995.

b.

Year	Price
1	$\$125,000 \times 1.07^1 = \$133,750$
2	$\$125,000 \times 1.07^2 = \$143,113$
3	$\$125,000 \times 1.07^3 = \$153,130$
4	$\$125,000 \times 1.07^4 = \$163,850$
5	$\$125,000 \times 1.07^5 = \$175,319$
6	$\$125,000 \times 1.07^6 = \$187,591$
7	$\$125,000 \times 1.07^7 = \$200,723$
8	$\$125,000 \times 1.07^8 = \$214,773$
9	$\$125,000 \times 1.07^9 = \$229,807$
10	$\$125,000 \times 1.07^{10} = \$245,894$

c.

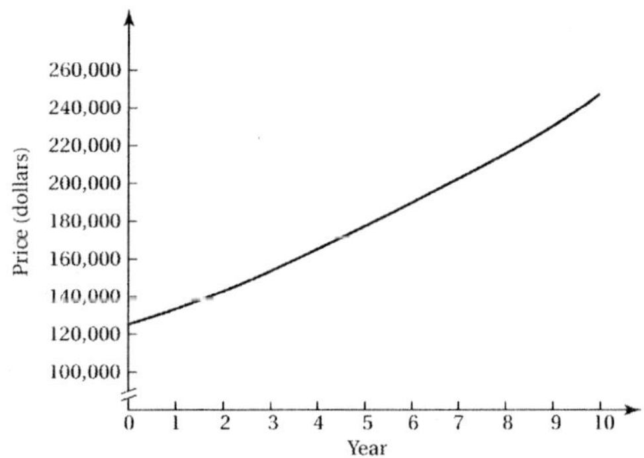

11. a. $Q = \$2000 \times 1.05^t$, where Q is your monthly salary, and t is the time (in years) since your salary was \$2000.

b.

Year	Monthly Salary
1	$\$2000 \times 1.05^1 = \$2100.00$
2	$\$2000 \times 1.05^2 = \$2205.00$
3	$\$2000 \times 1.05^3 = \$2315.25$
4	$\$2000 \times 1.05^4 = \$2431.01$
5	$\$2000 \times 1.05^5 = \$2552.56$
6	$\$2000 \times 1.05^6 = \$2680.19$
7	$\$2000 \times 1.05^7 = \$2814.20$
8	$\$2000 \times 1.05^8 = \$2954.91$
9	$\$2000 \times 1.05^9 = \$3102.66$
10	$\$2000 \times 1.05^{10} = \$3257.79$

c.

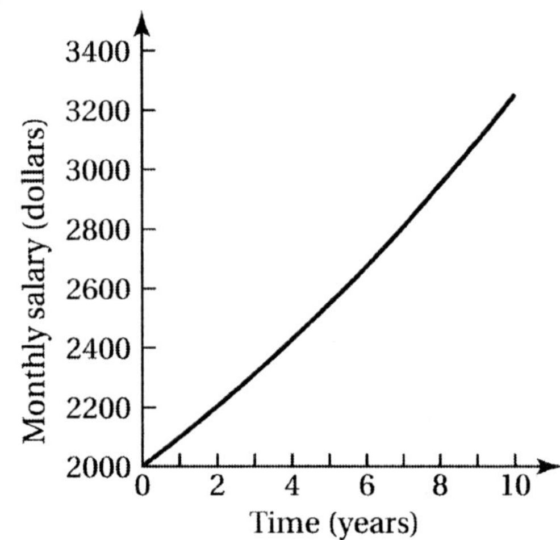

13. Metropolitan Population Growth. For an annual growth rate of 3%, a city with a 2000 population of 85,000 would have $85,000 \times 1.03^{10} = 114,233$ people by 2010.

The same city would have $85,000 \times 1.03^{100} = 1,633,584$ people by 2100.

Annual Versus Monthly Inflation.

15. A monthly inflation rate of 0.6% implies that after 12 months a price $\$Q$ will grow to $\$Q \times 1.006^{12} = \$Q \times 1.0744$, and so the annual inflation rate is 7.44%.

17. Hyperinflation in Brazil. A monthly inflation rate of 80% implies that after 12 months a price $\$Q$ will grow to $\$Q \times 1.80^{12} = \$Q \times 1156.83$, and so the annual inflation rate is 115,683%: things cost 1157 times what they cost a year earlier.

Since 1 day is 1/30 of a month, and after 1/30 of a month a price $\$Q$ will grow to $\$Q \times 1.80^{1/30} = \$Q \times 1.01979$, the daily inflation rate was 1.98%.

19. Extinction by Poaching. An annual decay rate of 10% implies that after t years a population of 1000 animals is reduced to $Q = 1000 \times 0.90^t$. We need to find for what value of t we get $Q = 30$, the level of extinction. Set $1000 \times 0.90^t = 30$ and solve for t.

Divide by 1000 so that $0.90^t = 0.03$, then take logarithms to get $\log_{10}(0.90^t) = \log_{10}(0.03)$, i.e., $x \log_{10}(0.90) = \log_{10}(0.03)$. Dividing by $\log_{10}(0.90)$ we arrive at

$$x = \frac{\log_{10}(0.03)}{\log_{10}(0.90)} = 33.28 \text{ years.}$$

21. Valium Metabolism. The amount of Valium present in the bloodstream (in milligrams) is given by

$$Q = 20 \times \left(\frac{1}{2}\right)^{\frac{t}{36}},$$

where t is the time in hours after midnight.

a. By noon, when $t = 12$, we have

$$Q = 20 \times \left(\frac{1}{2}\right)^{\frac{12}{36}} = 15.9 \text{ milligrams.}$$

b. We need to find the value of t for which $Q = 0.10 \times 20 = 2$ milligrams. So we set $20 \times (\frac{1}{2})^{\frac{t}{36}} = 2$, and solve for t.

First, divide by 20 and rewrite as $0.50^{\frac{t}{36}} = 0.1$. Take logarithms to get $\log_{10}(0.50^{\frac{t}{36}}) = \log_{10}(0.1)$, i.e., $\frac{t}{36} \log_{10}(0.50) = \log_{10}(0.1) =$

-1. Multiplying by 36, and then dividing by $\log_{10}(0.50)$ yields

$$t = \frac{-36}{\log_{10}(0.50)} = 119.59 \text{ hours.}$$

or about 120 hours.

Radioactive Dating.

23. The current amount Q of uranium-238 present in the rock is given in terms of the original amount Q_0 present via the formula:

$$Q = Q_0 \times \left(\frac{1}{2}\right)^{\frac{t}{4.5}},$$

where t is time in billions of years.

a. If 85% of the original uranium-238 remains at time t, this means that

$$0.85 \times Q_0 = Q_0 \times \left(\frac{1}{2}\right)^{\frac{t}{4.5}},$$

and so, dividing by Q_0 and switching sides,

$$\left(\frac{1}{2}\right)^{\frac{t}{4.5}} = 0.85.$$

We must solve this equation for t.

Taking logarithms we get $\log_{10}(0.50^{\frac{t}{4.5}}) = \log_{10}(0.85)$, i.e., $\frac{t}{4.5}\log_{10}(0.50) = \log_{10}(0.85)$. Multiplying by 4.5, and then dividing by $\log_{10}(0.50)$, we find that

$$t = \frac{4.5\log_{10}(0.85)}{\log_{10}(0.50)} = 1.06 \text{ billion years.}$$

b. We repeat the calculations done in (a) with 55% in place of 85%, i.e., we solve

$$\left(\frac{1}{2}\right)^{\frac{t}{4.5}} = 0.55.$$

for t.

Proceeding as in (a), we eventually get

$$t = \frac{4.5\ \log_{10}(0.55)}{\log_{10}(0.50)} = 3.88 \text{ billion years.}$$

25. The current density Q of the substance is given in terms of the density Q_0 of the substance 45 years ago via the formula:

$$Q = Q_0 \times \left(\frac{1}{2}\right)^{\frac{t}{20}},$$

where t is time in years, with $t = 0$ corresponding to 45 years ago.

a. If a density of 2 milligrams per square meter is detected today, this means that

$$2 = Q_0 \times \left(\frac{1}{2}\right)^{\frac{45}{20}},$$

and so,

$$Q_0 = \frac{2}{\left(\frac{1}{2}\right)^{\frac{45}{20}}} = 9.5 \text{ mg/m}^2.$$

Unit 8A Mathematics and Music

1. Octaves. To raise any tone by one octave, we double its frequency. So, starting with a tone with frequency 110 cps, we get a frequency of 220 cps for the tone one octave higher. We double the frequency again to go another octave higher, getting a frequency of 440 cps for the tone two octaves higher than the original, a frequency of 880 cps for the tone three octaves higher than the original, and a frequency of 1760 cps for the tone four octaves higher than the original.

3. Notes of a Scale. Each half-step on a 12-tone scale corresponds to an increase in frequency by a factor of $f = 1.05946$. Hence, $Q = 437 \times 1.05946^n$ gives the frequency of a tone n half-steps above the A above middle C. Rounding to the nearest number, we obtain the frequencies in the following table:

Note	Frequency (cps)
A	437
A#	$437 \times 1.05946^1 = 463$
B	$437 \times 1.05946^2 = 491$
C	$437 \times 1.05946^3 = 520$
C#	$437 \times 1.05946^4 = 551$
D	$437 \times 1.05946^5 = 583$
D#	$437 \times 1.05946^6 = 618$
E	$437 \times 1.05946^7 = 655$
F	$437 \times 1.05946^8 = 694$
F#	$437 \times 1.05946^9 = 735$
G	$437 \times 1.05946^{10} = 779$
G#	$437 \times 1.05946^{11} = 825$
A	$437 \times 1.05946^{12} = 874$

5. Exponential Growth and Scales. $Q = 260 \times 1.05946^n$ gives the frequency of a tone n half-steps above middle C.

a. If $n = 5$ we get $Q = 260 \times 1.05946^5 = 347$ cps.

b. Since a fifth corresponds to 7 half-steps, we set $n = 7$ and get $Q = 260 \times 1.05946^7 = 390$ cps.

c. Since a fourth corresponds to 5 half-steps, an octave and a fourth accounts for $12 + 5 = 17$ half-steps, so we set $n = 17$ and get $Q = 260 \times 1.05946^{17} = 694$ cps.

d. If $n = 36$ we get $Q = 260 \times 1.05946^{36} = 2080$ cps.

e. Four octaves and 3 half-steps accounts for $4 \times 12 + 3 = 51$ half-steps, so we set $n = 51$ and get $Q = 260 \times 1.05946^{51} = 4946$ cps.

7. Exponential Decay and Scales. To find the frequency of a note n half-steps above (respectively below) another note, we multiply (respectively divide) by a factor of 1.05946^n. Hence, the frequency of the note 5 half-steps below the note with a frequency of 437 cps is $437/(1.05946^5) = 327$ cps; and the frequency of the note 8 half-steps below the note with a frequency of 437 cps is $437/(1.05946^8) = 275$ cps.

9. Circle of Fifths.

a. Each half-step upward on the scale increases the frequency by a factor of $f = 2^{\frac{1}{12}} = 1.05946$. Raising a note by a fifth, or seven half-steps, increases the frequency by a factor of $2^{\frac{7}{12}} = 1.05946^7 = 1.49828$.

b. Raising a note by two fifths, or fourteen half-steps, increases the frequency by a factor of $2^{\frac{14}{12}} = 1.05946^{14} = 2.24483$.

c. The table below shows the notes in the complete circle of fifths. We need 12 steps of a fifth to complete the circle, taking us through seven octaves. Each interval of a fifth raises the frequency by a factor of 1.498.

Note	Frequency (cps)
C	260
G	$260 \times 1.498^1 = 389$
D	$260 \times 1.498^2 = 583$
A	$260 \times 1.498^3 = 874$
E	$260 \times 1.498^4 = 1309$
B	$260 \times 1.498^5 = 1961$
F#	$260 \times 1.498^6 = 2938$
C#	$260 \times 1.498^7 = 4401$
G#	$260 \times 1.498^8 = 6593$
D#	$260 \times 1.498^9 = 9876$
A#	$260 \times 1.498^{10} = 14,794$
F	$260 \times 1.498^{11} = 22,162$
C	$260 \times 1.498^{12} = 33,198$

d. The circle passes through 12 notes, and 7 octaves. Note that the frequency of the last tone is $1.498^{12} = 127.69$, or almost 128, times the frequency of the first tone; also $2^7 = 128$, so we passed upward through seven octaves, each corresponding to a doubling of frequency.

e. As noted in (d) above, the ratio in question is close to 128.

f. Raising a note by a fourth, or five half-steps, increases the frequency by a factor of $2^{\frac{5}{12}} = 1.05946^5 = 1.33482$. If we raise a tone by 12 $(2^{\frac{5}{12}})^{12} = 2^5$, so a circle of fourths extends over five octaves.

Unit 8B Perspective and Symmetry

1. Vanishing Points.

a. A vanishing point of the picture is that point at which the edges of the road appear to meet. This vanishing point is not the principal vanishing point as defined in the text; that one corresponds to all lines that are parallel in the real scene and perpendicular to the canvas. The road in this scene is not perpendicular to the canvas.

b.

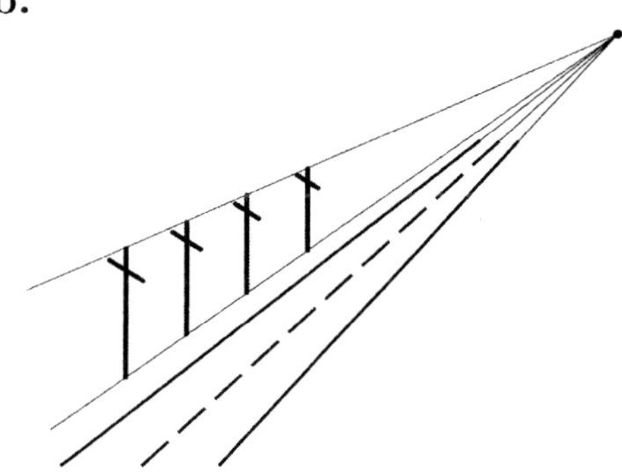

3. Drawing with Perspective.

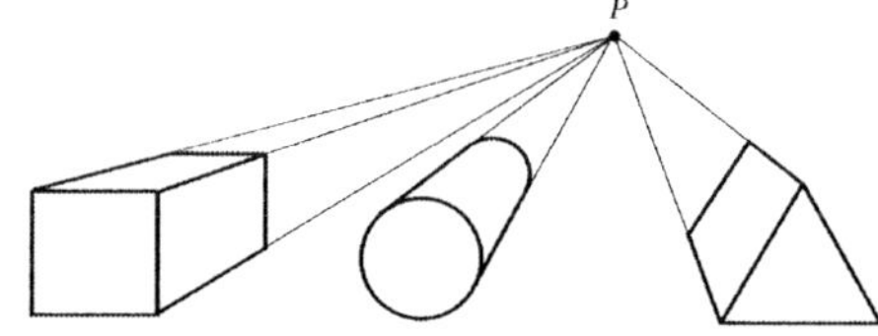

5. Proportion and Perspective.

a.

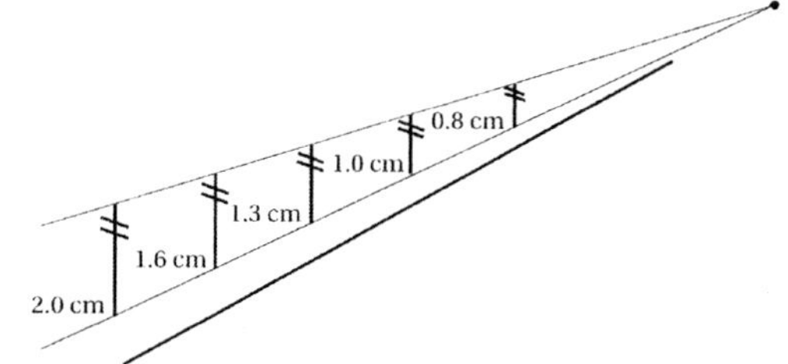

b. The ratios of the lengths of the two poles is $1.6/2.0 = 0.08$, and this ratio must be maintained for all successive poles spaced 2 cm apart. So the third pole has a height of $0.8 \times 1.6 = 1.28$ cm, the fourth pole has a height of $0.8 \times 1.28 = 1.024$ cm, and the fifth pole has a height of $0.8 \times 1.024 = 0.82$ cm.

c. No: if the poles are equally spaced in the drawing, then they would not be equally spaced in the real scene, or alternatively, if they are equally spaced in the real scene, then they would not appear to be equally spaced in the drawing.

7. Desargues' Theorem. See the figure given in the exercise.

9. Star Symmetries.

a. The four-pointed star has four reflection symmetries: about a vertical line, a horizontal line, and either of the two diagonal lines passing through the center.

It also has three rotational symmetries, corresponding to rotations of 90°, 180° or 270°.

b. The seven-pointed star has seven reflection symmetries: one for each line through one of the star points and the center.

It also has six rotational symmetries, corresponding to rotations of $\frac{360}{7}^o = 51.4286^o$ and multiples of this angle by factors of 2, 3, 4, 5 or 6.

Identifying Symmetries.

11. It has reflection symmetries: it can be reflected across a vertical line through its center, a horizontal line through its center, or either of its diagonals, and its appearance remains the same. It has rotational symmetries: it can be rotated though 90°, 180° and 270°, and its appearance remains the same.

13. It has translation symmetry (if imagined to be extended in both directions). It also has a reflection symmetry (across a horizontal line between the waves).

Tilings from Translating Triangles.

15.

Tilings from Translating and Reflecting Triangles.

17.

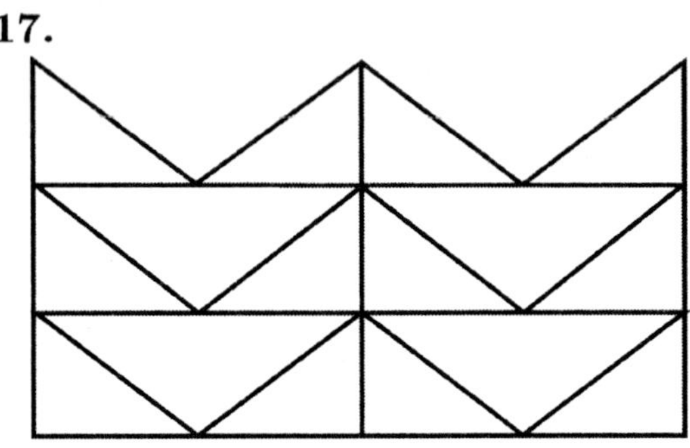

Tilings from Quadrilaterals.

19.

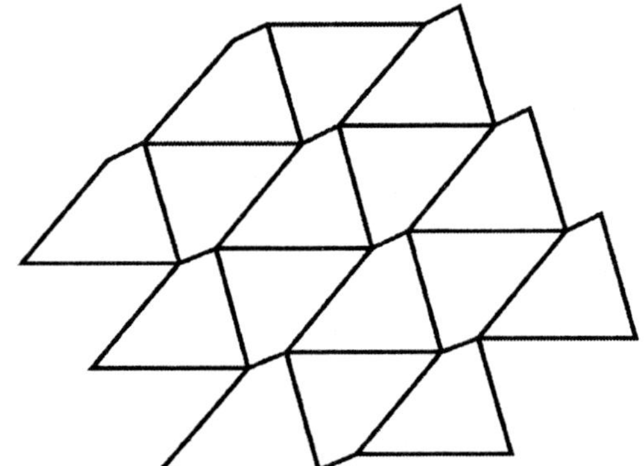

21. Why Quadrilateral Tilings Work. The angles around a point P are precisely the angles that appear inside of a single quadrilateral. Thus the angles around P have a sum of 360°, and the quadrilaterals around P fit together perfectly.

Unit 8C Proportion and the Golden Ratio

1. Golden Ratio. The longer segment should have length 2.47 inches, the shorter one 1.53 inches.

Dimensions of Golden Rectangles.

3. If the other side is longer, it's $1.5 \times \phi = 1.5 \times 1.62 = 2.43$ inches long; if it's shorter, it's $1.5/\phi = 1.5/1.62 = 0.93$ inches long.

5. If the other side is longer, it's $6.4 \times \phi = 6.4 \times 1.62 = 10.37$ km long; if it's shorter, it's $6.4/\phi = 6.4/1.62 = 3.95$ km long.

7. Solutions will vary.

9. Properties of ϕ.

a. $1/\phi = 1/1.618034 = 0.618034 = \phi - 1$.

b. $\phi^2 = 1.618034^2 = 2.618034 = \phi + 1$.

11. The Lucas Sequence.

a. Starting with $L_1 = 1$ and $L_2 = 3$, we get $L_3 = L_2 + L_1 = 3 + 1 = 4$, $L_4 = L_3 + L_2 = 4 + 3 = 7$, $L_5 = L_4 + L_3 = 11 + 7 = 18$, and so on, as tabulated below:

n	L_n	$\frac{L_n}{L_{n-1}}$
1	1	
2	3	3.000
3	4	1.333
4	7	1.750
5	11	1.571
6	18	1.636
7	29	1.611
8	47	1.621
9	76	1.617

b. The desired ratios have also been tabulated. They appear to be approaching the golden ratio ϕ.

Unit 9A Voting: Does the Majority Always Rule?

1. 1876 Presidential Election. To find a candidate's percent of the vote, divide the candidates votes by the total number of votes.

Total Popular Vote 8,430,783	Total Electoral Vote 369
Tilden 51.01%	Tilden 49.9%
Hayes 47.88%	Hayes 50.1%

Although Tilden narrowly won the popular vote, Hayes narrowly won the electoral vote and became President.

Other Close Presidential Elections.

3. To find a candidate's percent of the vote, divide the candidates votes by the total number of votes.

Total Popular Vote 8,891,088	Total Electoral Vote 369
Garfield 50.04%	Garfield 58.0%
Hancock 49.96%	Hancock 42.0%

Garfield narrowly won the popular vote and easily won the electoral vote to become President.

5. To find a candidate's percent of the vote, divide the candidates votes by the total number of votes.

Total Popular Vote 79,978,736	Total Electoral Vote 537
Carter 51.05%	Carter 55.3%
Ford 48.95%	Ford 44.7%

Carter narrowly won the popular vote and comfortably won the electoral vote to become President.

7. Super Majorities.

a. The percentage of shareholders favoring the merger is $10,580/15,890 = 66.58\%$ which is just shy of the required $2/3 = 66.67\%$ vote needed.

b. A 3/4 vote of a 15-member jury is at least 12 votes (since $0.75 \times 15 = 11.25$). Thus there will be no conviction.

c. A 3/4 super majority of the states is needed to amend the Constitution. In this case, 32 of the 50 states, or 64% of the states, support the amendment, so it fails to pass.

d. A $2/3 = 66.67\%$ vote is needed in both the House and the Senate to override a veto. The override gets a $68/100 = 68\%$ vote in the Senate and a $292/435 = 67.13\%$ vote in the House. So the veto can be overturned.

9. Three-Candidate Elections.

a. Best wins a plurality, but not a majority.

b. Imagine that there are 100 votes cast in the entire election. Then Able needs 18 votes to win 50% of the votes. Eighteen votes represents $18/28 = 64.29\%$ of Crown's votes.

11. Three-Candidate Elections.

a. The total number of votes cast is 465. Inviglio, who has the most votes, wins $185/465 = 39.78\%$ of the votes, which is a plurality, but not a majority.

b. A majority of the votes is 233 votes; so Height needs 83 votes to overtake Inviglio, which is $83/130 = 63.85\%$ of Grand's votes.

13. Ballots to Schedule. There are four different rankings of the three brands. Be sure the total votes add up to 10.

First	A	A	C	C
Second	B	C	A	B
Third	C	B	B	A
	4	2	3	1

Preference Schedules.

15. a. A total of 66 votes were cast.

b. A received 8 first place votes, B received 20 first place votes, C received 16 first place votes, and D received 22 first place votes. Therefore, D is the plurality winner (but not by a majority).

c. From part (b), we see that B and D enter the runoff and the votes of A and C are redistributed. Now B receives $20 + 6 = 26$ votes and D receives the remainder, or 40, of the votes. Thus D is the winner of the top two runoff.

d. In the sequential runoff we eliminate only the candidate with the fewest first place votes at each stage. From part (b), we see that A is eliminated first. Redistributing A's votes, D receives A's 8 first place votes; so at this point B has 20 votes, C has 16 votes, and D has 30 votes. Now C is eliminated and the election is between B and D, as in part (c). The winner by sequential runoff is D.

e. For the Borda Count, we score 4 points for a first place vote, 3 points for a second place vote, 2 points for a third place vote, and 1 point for a fourth place vote. The point totals are as follows:

$$\text{A: } (20 \times 1) + (15 \times 3) + (10 \times 2) + (8 \times 4) + (7 \times 3) + (6 \times 3) = 156.$$

$$\text{B: } (20 \times 4) + (15 \times 1) + (10 \times 1) + (8 \times 1) + (7 \times 2) + (6 \times 2) = 139.$$

$$\text{C: } (20 \times 2) + (15 \times 2) + (10 \times 4) + (8 \times 2) + (7 \times 1) + (6 \times 4) = 157.$$

$$\text{D: } (20 \times 3) + (15 \times 4) + (10 \times 3) + (8 \times 3) + (7 \times 4) + (6 \times 1) = 208.$$

Note that the point total is 660, as it must be. We see that D is the winner by the Borda count.

f. Here are the results of the 6 pairwise races:

A over B, 46 to 20.
C over A, 36 to 30.
D over A, 52 to 14.
C over B, 39 to 27.
D over B, 40 to 26.
D over C, 50 to 16.

Thus A scores 1 point, B scores 0 points, C scores 2 points and D scores 3 points. D wins by the pairwise comparison method.

g. As the winner by all five methods, candidate D is clearly the winner of the election.

17. a. A total of 100 votes were cast.

b. A received 35 first place votes, B received 25 first place votes, and C received 40 first place votes. Therefore, C is the plurality winner (but not by a majority).

c. From part (b), we see that A and C enter the runoff and the votes of B are redistributed. Now A receives 20 votes and C receives the remainder, or 5, of the votes. A now has 55 votes and C has 45 votes. Thus A is the winner of the top two runoff.

d. With only three candidates, the sequential runoff method is the same as the top two runoff method.

e. For the Borda Count, we score 3 points for a first place vote, 2 points for a second place vote, and 1 point for a third place vote. The point totals are as follows:

$$A: (30 \times 3) + (5 \times 3) + (20 \times 2) + (5 \times 1) + (10 \times 2) + (30 \times 1) = 200.$$
$$B: (30 \times 2) + (5 \times 1) + (20 \times 3) + (5 \times 3) + (10 \times 1) + (30 \times 2) = 210.$$
$$C: (30 \times 1) + (5 \times 2) + (20 \times 1) + (5 \times 2) + (10 \times 3) + (30 \times 3) = 190.$$

Note that the point total is 600, as it must be. We see that B is the winner by the Borda count.

f. Here are the results of the 6 pairwise races:

B over A, 55 to 45.
A over C, 55 to 45.
B over C, 55 to 45.

Thus A scores 1 point, B scores 2 points, and C scores no points. B wins by the pairwise comparison method.

g. A wins by runoff, B wins by Borda count and pairwise comparisons, C is the plurality winner. There is not a clear winner.

19. a. A total of 90 votes were cast.

b. A received no first place votes, B received 30 first place votes, C received no first place votes, D received 20 first place votes, and E received 40 first place votes. Therefore, E is the plurality winner (but not by a majority).

c. From part (b), we see that B and E enter the runoff and the votes of A, C, and D are redistributed. Now B picks up 20 votes, making B the winner of the top two runoff.

d. Since only three candidates received first place votes, the sequential runoff and the top two runoff methods are the same.

e. For the Borda Count, we score 5 points for a first place vote, 4 points for a second place vote, 3 points for a third place vote, 2 points for a fourth place vote, and 1 point for a fifth place vote. The point totals are as follows:

$$A: (40 \times 3) + (30 \times 2) + (20 \times 4) = 260.$$
$$B: (40 \times 2) + (30 \times 5) + (20 \times 3) = 290.$$
$$C: (40 \times 1) + (30 \times 4) + (20 \times 2) = 200.$$
$$D: (40 \times 4) + (30 \times 1) + (20 \times 5) = 290.$$
$$E: (40 \times 5) + (30 \times 3) + (20 \times 1) = 310.$$

Note that the point total is 1350, as it must be. We see that E is the winner by the Borda count.

f. Here are the results of the 10 pairwise races:

A over B, 60 to 30.
A over C, 60 to 30.
D over A, 60 to 30.
E over A, 70 to 20.
B over C, 90 to 0.
D over B, 60 to 30.
B over E, 50 to 40.
D over C, 60 to 30.
C over E, 50 to 40.
E over D, 70 to 20.

Thus A scores 2 points, B scores 2 points, C scores 1 point, D scores 3 points, and E scores 2 points. By the pairwise comparison method, D is the winner.

g. Candidates B and E each win two of the five methods, so the outcome is debatable.

21. Pairwise Comparison Question. With four candidates, there are $3+2+1=6$

pairwise races. With five candidates, there are $4+3+2+1=10$ pairwise races.

23. Borda Question. Each voter will award a first, second, third and fourth place vote. The points for these places total $4+3+2+1=10$ points. If there are 25 voters, there are $25 \times 10 = 250$ points. Candidates A, B, and C have a total of 120 points, so candidate D must have 130 points which is enough to win the election.

Unit 9B Apportionment: The House of Representatives and Beyond

State Representation. Recall that

$$\text{standard divisor} = \frac{281 \text{ million}}{435} = 646,000,$$

and

$$\text{standard quota} = \frac{\text{state population}}{\text{standard divisor}}.$$

1. For Connecticut, we have

$$\text{standard quota} = \frac{3,406,000}{646,000} = 5.27,$$

which is bigger than the 5 seats the state actually has, so Connecticut is underrepresented in the House.

3. For Florida, we have

$$\text{standard quota} = \frac{15,982,000}{646,000} = 24.74,$$

which is smaller than the 25 seats the state actually has, so Florida is overrepresented in the House.

Alabama Paradox.

5. The total population is $950+670+240 = 1866$, so with 100 seats to be apportioned, the standard divisor is $1866/100 = 18.66$. This is used to compute the standard quota, and hence the minimum quotas, in the table below. Hamilton's method applied to these three states yields:

State	A	B	C	Total
Pop.	950	670	246	1866
Std. Q.	50.91	35.91	13.18	100
Min. Q.	50	35	13	98
Frac. R.	0.91	0.91	0.18	2
Final A.	51	36	13	100

Assuming 101 delegates, Hamilton's method applied to these three states yields:

State	A	B	C	Total
Pop.	950	670	246	1866
Std. Q.	51.42	36.26	13.32	101
Min. Q.	51	36	13	100
Frac. R.	0.42	0.26	0.32	1
Final A.	52	36	13	101

No state lost seats as a result of the additional available seat, so the Alabama paradox does not occur here.

7. The total population is $770+155+70+673 = 1668$, so with 100 seats to be apportioned, the standard divisor is $1668/100 = 16.68$. This is used to compute the standard quota, and hence the minimum quotas, in the table below. Hamilton's method applied to these four states yields:

State	A	B	C	D	Total
Pop.	770	155	70	673	1668
Std. Q.	46.16	9.29	4.20	40.35	100
Min. Q.	46	9	4	40	99
Frac. R.	0.16	0.29	0.20	0.35	1
Final A.	46	9	4	41	100

Assuming 101 delegates, Hamilton's method applied to these four states yields:

State	A	B	C	D	Total
Pop.	770	155	70	673	1668
Std. Q.	46.62	9.39	4.24	40.75	101
Min. Q.	46	9	4	40	99
Frac. R.	0.62	0.39	0.24	0.75	2
Final A.	47	9	4	41	101

No state lost seats as a result of the additional available seat, so the Alabama paradox does not occur here.

New States Paradox.

9. The total population is $1140 + 6320 + 250 = 7710$, so with 100 seats to be apportioned, the standard divisor is $7710/100 = 77.10$. This is used to compute the standard quota, and hence the minimum quotas, in the table below. Hamilton's method applied to these three states yields:

State	A	B	C	Total
Pop.	1140	6320	250	7710
Std. Q.	14.79	81.97	3.24	100
Min. Q.	14	81	3	98
Frac. R.	0.79	0.97	0.24	2
Final A.	15	82	3	100

With the addition of a new state D with population 500, for whom 5 new delegates are added, Hamilton's method applied to these four states yields:

State	A	B	C	D	Total
Pop.	1140	6320	250	500	8210
Std. Q.	14.58	80.83	3.20	6.39	105
Min. Q.	14	80	3	6	103
Frac. R.	0.58	0.83	0.20	0.39	2
Final A.	15	81	3	6	105

Even though 5 new seats were added with state D representation in mind, Hamilton's method actually assigned 6 seats to this state, 1 of them at the expense of State B. Since B lost a seat as a result of the additional seats for the new state, the New States paradox does occur here.

Jefferson's Method.

11. The total population is $98+689+212 = 999$, so with 100 seats to be apportioned, the standard divisor is $999/100 = 9.99$. This is used to compute the standard quota, and hence the minimum quotas, in the table below. Using a modified divisor of 9.83 instead, we get the modified quotas listed, and the new minimum quotas.

State	A	B	C	Total
Pop.	98	689	212	999
Std. Q.	9.81	68.97	21.22	100
Min. Q.	9	68	21	98
Mod. Q.	9.97	70.09	21.57	101.63
N. Min. Q.	9	70	21	100

Since the new minimum quota successfully apportions all 100 seats, we can stop. Note, however, that the quota criterion is violated, because State B's standard quota is 68.97 yet it ends up being apportioned 70 seats.

13. The total population is 979, so with 100 seats to be apportioned, the standard divisor is $979/100 = 9.79$. This is used to compute the standard quota, and hence the minimum quotas, in the table below. Using a modified divisor of 9.57 instead, we get the modified quotas listed, and the new minimum quotas.

State	A	B	C	D	Total
Pop.	69	680	155	75	979
Std. Q.	7.05	69.46	15.83	7.66	100
Min. Q.	7	69	15	7	98
Mod. Q.	7.19	70.83	16.15	7.81	101.98
N. Min. Q.	7	70	16	7	100

Since the new minimum quota successfully apportions all 100 seats, we can stop. The quota criterion is satisfied.

Comparing Methods.

15. a. The total population is $535 + 344 + 120 = 999$, so with 100 seats to be apportioned, the standard divisor is $999/100 = 9.99$. This is used to compute the standard quota, and hence the minimum quotas, in the table below. Hamilton's method applied to these three states yields:

State	A	B	C	Total
Pop.	535	344	120	999
Std. Q.	53.55	34.43	12.01	99.9
Min. Q.	53	34	12	99
Frac. R.	0.55	0.43	0.01	1
Final A.	54	34	12	100

b. Jefferson's method starts off like Hamilton's. As noted in (a), the standard divisor is 9.99, so we try lower modified divisors until the apportionment comes out just right. By trial and error, the modified divisor 9.90 is found to work, as documented in the last two rows of the table below. Note that this choice of modified divisor is not unique, other nearby values also work.

State	A	B	C	Total
Pop.	535	344	120	999
Std. Q.	53.55	34.43	12.01	99.9
Min. Q.	53	34	12	99
Mod. Q.	54.04	34.75	12.12	100.91
N. Min. Q.	54	34	12	100

c. Webster's method here requires us to find a modified divisor such that the corresponding modified quotas *round* (not truncate) to numbers which sum to the desired 100. Inspecting the tables in (a) or (b), we see that the standard divisor and standard quote are already adequate, as documented in the last row of the table below. Note that this choice of modified divisor is not unique, other nearby values also work.

State	A	B	C	Total
Pop.	535	344	120	999
Std. Q.	53.55	34.43	12.01	100
Min. Q.	53	34	12	99
Rou. Q.	54	34	12	100

d. The Hill-Huntington method here requires us to find a modified divisor such that the corresponding modified quotas *rounded relative to the geometric mean* yield numbers which sum to the desired 100. The standard divisor 9.99 fits the bill admirably, as documented in the last two rows of the table below. Note that this choice of divisor is not unique, other nearby values also work.

State	A	B	C	Total
Pop.	535	344	120	999
Std. Q.	53.55	34.43	12.01	100
Min. Q.	53	34	12	99
Geom. M.	53.50	34.50	12.49	-
Rou. Q.	54	34	12	100

The Geometric Means are for the whole numbers bracketing the standard quotas, namely, $\sqrt{53 \times 54} = 53.50$, $\sqrt{34 \times 35} = 34.50$, and $\sqrt{12 \times 13} = 12.49$. The modified quotas—in this case the standard quota—are compared to these, and hence 34.43 and 12.01 are rounded down to 34 and 12, respectively, whereas 53.55 is rounded up to

43. All told, for this example, the Hill-Huntington method worked out exactly the same as Webster's.

e. All four methods gave the same results.

17. a. The total population is $836+2703+2626+3835 = 10,000$, so with 100 seats to be apportioned, the standard divisor is $10,000/100 = 100$. This is used to compute the standard quota, and hence the minimum quotas, in the table below. Hamilton's method applied to these four states yields:

State	A	B	C	D	Total
Pop.	836	2703	2626	3835	10,000
Std. Q.	8.36	27.03	26.26	38.35	100
Min. Q.	8	27	26	38	99
Frac. R.	0.36	0.03	0.26	0.35	1
Final A.	9	27	26	38	100

b. Jefferson's method starts off like Hamilton's. As noted in (a), the standard divisor is 100, so we try lower modified divisors until the apportionment comes out just right. By trial and error, the modified divisor 98.3 is found to work, as documented in the last two rows of the table below. Note that this choice of modified divisor is not unique, other nearby values also work.

State	A	B	C	D	Total
Pop.	836	2703	2626	3835	10,000
Std. Q.	8.36	27.03	26.26	38.35	100
Min. Q.	8	27	26	38	99
Mod. Q.	8.50	27.50	26.71	39.01	101.73
N. Min. Q.	8	27	26	39	100

c. Webster's method here requires us to find a modified divisor such that the corresponding modified quotas *round* (not truncate) to numbers which sum to the desired 100. Inspecting the tables in (a) or (b), we see that neither the standard nor modified divisors and quotas there work, so we must try other modified divisors. By trial and error, we find that 99.5 is a suitable modified divisor, as documented in the last rows of the table below. Note that this choice of modified divisor is not unique, other nearby values also work.

State	A	B	C	D	Total
Pop.	836	2703	2626	3835	10,000
Std. Q	8.36	27.03	26.26	38.35	100
Min. Q.	8	27	26	38	99
Mod. Q.	8.40	27.17	26.39	38.54	100.50
Rou. Q.	8	27	26	39	100

d. The Hill-Huntington method here requires us to find a modified divisor such that the corresponding modified quotas *rounded relative to the geometric mean* yield numbers which sum to the desired 100. The standard divisor does not work here, nor does the modified divisor of (b), but the modified divisor 99.5 encountered in (c) turns out to be suitable, as documented in the last three rows of the table below. Note that this choice of divisor is not unique, other nearby values also work.

State	A	B	C	D	Total
Pop.	836	2703	2626	3835	10,000
Std. Q	8.36	27.03	26.26	38.35	100
Min. Q.	8	27	26	38	99
Mod. Q.	8.40	27.17	26.39	38.53	100.49
Geom. M.	8.49	27.50	26.50	38.50	-
Rou. Q.	8	27	26	39	100

The Geometric Means are for the whole numbers bracketing the modified quotas, namely, $\sqrt{8\times 9} = 8.49$, $\sqrt{27\times 28} = 27.50$,

$\sqrt{26 \times 27} = 26.50$ and $\sqrt{38 \times 39} = 38.50$. The modified quotas are compared to these, and hence 8.40, 27.17 and 26.39 are rounded down to 8, 27 and 26, respectively, whereas 38.53 is rounded up to 39. All told, for this example, the Hill-Huntington method worked out exactly the same as Webster's.

e. The Jefferson, Webster and Hill-Huntington methods all gave the same result, so one could argue that these yield the best apportionment.

19. a. The total population is $48 + 97 + 245 = 390$, so with 10 committee positions to be apportioned, the standard divisor is $390/10 = 39$. This is used to compute the standard quota, and hence the minimum quotas, in the table below. Hamilton's method applied to these three groups yields:

Group	Soc.	Pol.	Ath.	Total
Pop.	48	97	245	390
Std. Q.	1.23	2.49	6.28	10
Min. Q.	1	2	6	9
Frac. R.	0.23	0.49	0.28	1
Final A.	1	3	6	10

b. Jefferson's method starts off like Hamilton's. As noted in (a), the standard divisor is 39, so we try lower modified divisors until the apportionment comes out just right. By trial and error, the modified divisor 35 is found to work, as documented in the last two rows of the table below. Note that this choice of modified divisor is not unique, other nearby values also work.

Group	Soc.	Pol.	Ath.	Total
Pop.	48	97	245	390
Std. Q.	1.23	2.49	6.28	10
Min. Q.	1	2	6	9
Mod. Q.	1.37	2.77	7.00	11.14
N. Min. Q.	1	2	7	10

c. Webster's method here requires us to find a modified divisor such that the corresponding modified quotas *round* (not truncate) to numbers which sum to the desired 10. Inspecting the tables in (a) or (b), we see that neither the standard nor modified divisors and quotas there work, so we must try other modified divisors. This time, we find that 38 is a suitable modified divisor, as documented in the last rows of the table below. Note that this choice of modified divisor is not unique, other nearby values also work.

Group	Soc.	Pol.	Ath.	Total
Pop.	48	97	245	390
Std. Q.	1.23	2.49	6.28	10
Min. Q.	1	2	6	9
Mod. Q.	1.26	2.55	6.45	10.26
Rou. Q.	1	3	6	10

d. The Hill-Huntington method here requires us to find a modified divisor such that the corresponding modified quotas *rounded relative to the geometric mean* yield numbers which sum to the desired 10. The standard divisor 39 fits the bill admirably, as documented in the last two rows of the table below. Note that this choice of divisor is not unique, other nearby values also work.

Group	Soc.	Pol.	Ath.	Total
Pop.	48	97	245	390
Std. Q.	1.23	2.49	6.28	10
Min. Q.	1	2	6	9
Geom. M.	1.41	2.45	6.48	-
Rou. Q.	1	3	6	10

The Geometric Means are for the whole numbers bracketing the standard quotas, namely, $\sqrt{1 \times 2} = 1.41$, $\sqrt{2 \times 3} = 2.45$, and $\sqrt{6 \times 7} = 6.48$. The modified quotas are compared to these, and hence 1.23 and 6.28 are rounded down to 1 and 6, respectively, whereas 2.49 is rounded up to 3. All told, for this example, the Hill-Huntington method worked out exactly the same as Webster's.

e. The Hamilton, Webster and Hill-Huntington methods all gave the same result, so one could argue that these yield the best apportionment.

21. a. The total "population" here is the total monthly gross sales, i.e., $2.5+7.6+3.9+5.5 = 19.5$ (in millions of dollars), so with 25 managers to be apportioned according to these sales, the standard divisor is $19.5/25 = 0.78$. This is used to compute the standard quota, and hence the minimum quotas, in the table below. Hamilton's method applied to these four stores yields:

Store	Bo.	D.	Br.	Ft. Co.	Total
Pop.	2.5	7.6	3.9	5.5	19.5
Std. Q.	3.21	9.74	5.00	7.05	25
Min. Q.	3	9	5	7	24
Frac. R.	0.21	0.74	0.00	0.05	1
Final A.	3	10	5	7	25

b. Jefferson's method starts off like Hamilton's. As noted in (a), the standard divisor is 0.78, so we try lower modified divisors until the apportionment comes out just right. By trial and error, the modified divisor 0.76 is found to work, as documented in the last two rows of the table below. Note that this choice of modified divisor is not unique, other nearby values also work.

Store	Bo.	D.	Br.	Ft. Co.	Total
Pop.	2.5	7.6	3.9	5.5	19.5
Std. Q.	3.21	9.74	5.00	7.05	25
Min. Q.	3	9	5	7	24
Mod. Q.	3.29	10.00	5.13	7.24	25.66
N. Min. Q.	3	10	5	7	25

c. Webster's method here requires us to find a modified divisor such that the corresponding modified quotas *round* (not truncate) to numbers which sum to the desired 25. Inspecting the tables in (a) or (b), we see that the standard divisor and standard quota are already adequate, as documented in the last row of the table below. Note that this choice of modified divisor is not unique, other nearby values also work.

Store	Bo.	D.	Br.	Ft. Co.	Total
Pop.	2.5	7.6	3.9	5.5	19.5
Std. Q.	3.21	9.74	5.00	7.05	25
Min. Q.	3	9	5	7	24
Rou. Q.	3	10	5	7	25

d. The Hill-Huntington method here requires us to find a modified divisor such that the corresponding modified quotas *rounded relative to the geometric mean* yield numbers which sum to the desired 25. The standard divisor 0.78 fits the bill admirably, as documented in the last two rows of the table below. Note that this choice of divisor is not unique, other nearby values also work.

Store	Bo.	D.	Br.	Ft. Co.	Total
Pop.	2.5	7.6	3.9	5.5	19.5
Std. Q.	3.21	9.74	5.00	7.05	25
Min. Q.	3	9	5	7	24
Geom. M.	3.46	9.49	5.48	7.48	-
Rou. Q.	3	10	5	7	25

The Geometric Means are for the whole numbers bracketing the standard quotas, namely, $\sqrt{3 \times 4} = 3.46$, $\sqrt{9 \times 10} = 9.49$, $\sqrt{5 \times 6} = 5.48$ and $\sqrt{7 \times 8} = 7.48$. The standard quotas are compared to these, and hence 3.21, 5.00 and 7.05 are rounded down to 3, 5 and 7, respectively, whereas 9.74 is rounded up to 10. All told, for this example, the Hill-Huntington method worked out exactly the same as Webster's.

e. All four methods gave the same results.